Reindeer on South Georgia

Studies in Polar Research

This series of publications reflects the growth of research activity in and about the polar regions, and provides a means of disseminating the results. Coverage is international and interdisciplinary: the books will be relatively short and fully illustrated. The intention of the series is to provide surveys of the present state of knowledge in a given subject, rather than research reports, conference proceedings or collected papers. The scope of the series is wide and includes studies in all the biological, physical and social sciences.

Editorial Board

A. Clarke, British Antarctic Survey, Cambridge
D. J. Drewry, Cambridge
S. W. Greene, Department of Botany, University of Reading
P. Wadhams, Scott Polar Research Institute, Cambridge
D. W. H. Walton, British Antarctic Survey, Cambridge
P. J. Williams, Geotechnical Research Laboratories Carleton University, Ottawa

Other titles in this series:

The Antarctic Circumpolar Ocean
Sir George Deacon

The Living Tundra*
Yu. I. Chernov, transl. D. Love

Arctic Air Pollution
edited by B. Stonehouse

The Antarctic Treaty Regime
edited by Gillian D. Triggs

Antarctica: The Next Decade
edited by Sir Anthony Parsons

Antarctic Mineral Exploitation
Francisco Orrego Vicuna

Transit Management in the Northwest Passage
edited by C. Lamson and D. VanderZwaag

Canada's Arctic Waters in International Law
Donat Pharand

Vegetation of the Soviet Polar Deserts
V. Aleksandrova, transl. D. Love

* Also issued as a paperback

Reindeer in front of Neumayer Glacier and the Allardyce Range on South Georgia

REINDEER
ON SOUTH GEORGIA

The ecology of an introduced population

N. LEADER-WILLIAMS

British Antarctic Survey, Cambridge

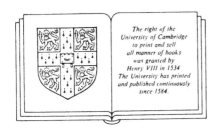

The right of the
University of Cambridge
to print and sell
all manner of books
was granted by
Henry VIII in 1534
The University has printed
and published continuously
since 1584.

CAMBRIDGE UNIVERSITY PRESS

Cambridge

New York   New Rochelle   Melbourne   Sydney

CAMBRIDGE UNIVERSITY PRESS
Cambridge, New York, Melbourne, Madrid, Cape Town, Singapore, São Paulo, Delhi

Cambridge University Press
The Edinburgh Building, Cambridge CB2 8RU, UK

Published in the United States of America by Cambridge University Press, New York

www.cambridge.org
Information on this title: www.cambridge.org/9780521106986

First published 1988
This digitally printed version 2009

A catalogue record for this publication is available from the British Library

Library of Congress Cataloguing in Publication data
Leader-Williams, N.
Reindeer on South Georgia: the ecology of an introduced population N.
Leader-Williams.
p. cm. – (Studies in polar research)
Bibliography: p.
Includes index.
ISBN 0 521 24271 1
1. Reindeer – South Georgia Island.   2. Reindeer – South Georgia
Island – Ecology.   3. Reindeer – Antarctic regions.   4. Reindeer–
Antarctic regions – Ecology.   5. Animal introduction – South Georgia
Island.   6. Animal introduction – Antarctic regions.   7. Reindeer–
Arctic regions.   8. Mammals – South Georgia Island.   9. Mammals–
South Georgia Island – Ecology.   10. Mammals – Antarctic
regions.   11. Mammals – Antarctic regions – Ecology.   12. Mammals
– Arctic regions. I. Title. II. Series.
QL737.U55L42 1988
599.73''57–dc 19   87-25060
                                                                          CIP

ISBN 978-0-521-24271-4 hardback
ISBN 978-0-521-10698-6 paperback

For my father and in memory of my mother

Contents

Contents   ix

Preface

The faunas and floras of Arctic and Antarctic regions are both adapted to the climatic extreme of cold, but differ in their evolution and ecology (Dunbar, 1977). The Arctic is an ice-covered ocean basin mostly encircled by continental or large island land masses, where temperatures can rise to well above freezing in summer. Although extensively glaciated on several occasions during the last one million years, Arctic tundra now supports simple but well-developed floras and faunas which were able to survive and evolve further in glacial refugia, and recolonise during inter-glacial periods. In contrast, the Antarctic continent is colder, has remained frozen for over 20 million years and supports the most limited terrestrial flora and fauna of any continent, and terrestrial life only survives close to continental margins. Antarctic tundra habitats have developed only on oceanic islands, but are poor in species due to the extreme isolation of each island group from other continental land masses. Antarctic animal life is essentially marine and is based largely upon krill which is at the centre of the food-web of whales, seals and seabirds in the Southern Ocean.

Reindeer and caribou (*Rangifer tarandus*) are a numerically prominent and sociologically important species amongst the Arctic fauna. Their common western European name is derived from the Lapp name of 'reino' for young of the species, whereas the common North American name of caribou derives from the Micmac Indian word 'xalibu', meaning 'pawer' or 'shoveller' in reference to their method of obtaining food through snow in winter (Banfield, 1961). They occur at higher latitudes than other species of deer and have adapted to the rigours of life in the Arctic tundra. As a species, reindeer and caribou are generally noted for their exceptionally long migrations, for their large group sizes, for their dependence on lichens as forage in winter and for providing food for wolves and man. The introduction of Norwegian reindeer to the subantarctic island of South Georgia earlier this century resulted in an unusual interaction between this Arctic mammal and a southern ecosystem. Early studies (Olstad, 1930; Bonner, 1958) showed that reindeer had adapted to an unusual diet and had increased greatly in numbers from the few animals originally introduced. In this monograph I will develop these observations into two main themes.

First, this is a comparative study of the biology and ecology of a cervid which occurs in a wide variety of habitats throughout the Arctic. Comparative studies amongst other groups of mammals (primates: Crook & Gartlan, 1966; Clutton-Brock, 1974; sheep: Geist, 1971; antelopes: Jarman, 1974; seals, ungulates and primates: Alexander *et al.*, 1979) have explained many differences in behaviour and ecology between and within species, and their particular adaptations to their environments. The study on South Georgia provides a basis for an intraspecific comparison between this unusual population and those occurring on natural range elsewhere. Additionally, studies of other cervids (e.g. Clutton-Brock, Albon & Harvey, 1980; Clutton-Brock, Guinness & Albon, 1982) have shown that characteristics favouring polygyny, such as large body size in males relative to females and large weapon size in males, are more enhanced in species living in large groups such as reindeer. Such dimorphism may lead to interesting sexual differences in feeding habits, growth and reproduction. Therefore, the adaptations shown by the most northerly species of cervid are investigated in interspecific comparisons with other temperate species.

The second theme of this study concerns the ways in which an introduced population adapts to a new environment, and its relevance to the population ecology of mammals and to the conservation of southern island ecosystems. The size of natural populations may fluctuate irregularly between restricted limits (Lack, 1954), but little has been learned of population processes in such situations (Caughley, 1976a). Much has been learned of population processes and of the ungulate–vegetation interaction in situations of greater instability. More dramatic fluctuations in population numbers amongst deer and other ungulates occur as a result of human interference. Increases result from predator removal (including the cessation of heavy hunting by man) or the introduction and subsequent eradication of a disease, whilst declines arise from the overutilisation of resources (Rasmussen, 1941; Murie, 1944; Jordan, Botkin & Wolfe, 1971; Bergerud, 1974c; Sinclair, 1979; Owen-Smith, 1981; Clutton-Brock *et al.*, 1982; Pellew, 1983). However, the most dramatic fluctuations of all arise when an ungulate encounters a new habitat, such as occurred with the introduction of Himalayan thar to New Zealand or of reindeer to Arctic islands (Caughley, 1976a,b). The reindeer on South Georgia provide an opportunity to study introductions made under comparable circumstances to those on Arctic islands. They also provide the opportunity to assess the impact of an introduced species upon a southern island ecosystem that is not adapted to grazing (Holdgate & Wace, 1961; Clark & Dingwall, 1985).

The monograph is arranged into four parts. Part I is a preamble to the study, Part II discusses the comparative biology of reindeer, whilst Part III discusses the consequences of the introduction of reindeer to South Georgia. Part IV provides a short overview of the study. Readers intent on methods and statistical justification of the results will need to include the information printed in small type in their reading of the monograph, whilst those interested in the results themselves may pass over these paragraphs. Much of the work described in this monograph has been published already in separate articles in a variety of specialist journals, each catering for its own readership. The invitation from Cambridge University Press to prepare this monograph gave me the opportunity to bring the various threads of the study together in one volume aimed at a wider readership. This study was undertaken some time ago and the delay in the appearance of this monograph resulted from my involvement with another field project in Africa. Nonetheless, I hope this contribution will add to our knowledge and appreciation of polar animals and cold regions.

Acknowledgements

It is a great pleasure to thank the many people who have helped with this study over the years. My greatest single debt is to the British Antarctic Survey and its staff, both on South Georgia and in Cambridge. I am particularly grateful to the then Director, Richard Laws, for his continuous encouragement.

On South Georgia, base members and ships' crews all helped at one time or another with fieldwork and logistics. Notable amongst these were Mike Payne and Peter Prince who undertook a preliminary project and Jim Boyle, Dave Burkitt, John Hall and, most especially, Bob Pratt who provided assistance and companionship in the field. Helicopter crews from HMS *Endurance* kindly undertook aerial counts. Simon Kightley, Nigel Lowthrop, Bob Headland and Tim Heilbronn recorded data and maintained experimental exclosure sites.

In Cambridge, Bob Pratt, Liz Kirkwood, Tessa Scott, Nigel Bonner, Howard Platt, Mike Payne, David Walton, Roger Worland and Alison Rosser assisted with laboratory material. Hormone analyses were undertaken at the MRC Unit of Reproductive Biology, Edinburgh with the help of Roger Short and Gerald Lincoln; histological analysis of reproductive material was completed at the AFRC Institute of Animal Physiology, Cambridge with the help of Hector Dott; rumen samples were milled at the University of Aston and ruminal nitrogen levels were measured at Merlewood Research Station, Institute of Terrestrial Ecology; parasites were identified by Roger Connan of the University of Cambridge and by Lynda Gibbons and Arlene Jones of the Commonwealth Institute of Helminthology; pathological material was examined by Arthur Jennings of the University of Cambridge and by Gwynneth Lewis and Carol Richardson of the Central Veterinary Laboratory. Steve Albon, Alastair Murray, Peter Rothery and Chris Ricketts have all helped greatly with, or collaborated in, data analysis.

The book has benefited greatly from the constructive criticism of various colleagues. Individual chapters were read by Gerald Lincoln, Ron Lewis Smith, Peter Rothery, John Croxall and Nigel Bonner. I am especially grateful to Steve Albon, Iain Gordon and David Walton for reviewing the whole manuscript. Sue Murray, Tony Sylvester and Chris Gilbert prepared the illustrations with great care, and the burden of typing and retyping the manuscript fell to Sue Norris, which she executed with skill and cheerfulness.

The editorial board of *Studies in Polar Research* and Cambridge University Press have been very patient in awaiting final delivery of the manuscript whilst I studied black rhinos in Zambia. David Walton in particular, has provided the necessary encouragement to complete the book, which Jane King subedited with great care. The support of my family has been very important, that of my parents to disappear to South Georgia in the first instance, and of Alison whilst writing the book.

For permission to use their photographs, I am grateful to David Klein, Nick Tyler, Dan Roby, Geoff Copson, Colin Meurk, Phillipe Vernon, and the New Zealand Forest Service. The British Antarctic Survey provided most of the funds and facilities for this project, including the preparation of all the illustrations and the final manuscript. A generous grant from the Falkland Islands Dependencies Fund (FIDF) allowed me to complete the book and covered the cost of the colour frontispiece, whilst a further grant from FIDF together with a grant from the Binks Trust has helped, by reducing the cost of production, to make the book available at its present price.

Part I

Reindeer in the Arctic and Antarctic

This section is an introductory preamble. It aims to set the scene for readers not familiar with reindeer, South Georgia or the consequences of introducing mammals to new environments. Reindeer and caribou are a species of northern herbivore uniquely adapted to Arctic life. They fill the lichen-based food niche and are predated upon by wolves. Man now has a pervasive effect on the fate of most reindeer and caribou populations (Chapter 1). Included amongst man's activities is the introduction of mammals to novel environments. This may result in dramatic irruptions, and introductions of reindeer made to islands are of considerable comparative interest (Chapter 2). Amongst recent introductions has been the establishment of northern herbivores on southern islands.
South Georgia lies in the subantarctic and lacks indigenous terrestrial mammals. The island's fauna is dominated by marine species and its terrestrial ecosystems are simple. Reindeer were introduced early this century and have had a considerable impact on plant communities, which led to the initiation of this study (Chapter 3).

Figure 1.1 Reindeer and caribou are unusual amongst cervids in that both males and females possess antlers. Antlers are used as a taxonomic feature for the genus *Rangifer*.

1

Reindeer and caribou: taxonomy, habitats and ecology

Tarandus. Horns branched, round, recurvate; summits palmate.　　*Rein Deer* Inhabits the Alpine mountains of *America*, *Europe* and *Asia*, southern parts of *Russia* and *Sardinia*; descends in winter into the plains, and is driven back into the mountains in summer by the persecution of marsh insects: feeds on the rein-deer lichen, which in winter it digs out of the snow with its feet; the male casts his horns at the end of November, the female not till she fawns, about the middle of May; gravid 33 weeks, brings often twins; lives about 16 years; when castrated loses the horns, not till the 9th year: is trained in Lapland to draw sledges, and supplies the inhabitants with milk, flesh and clothing; when domesticated 3 feet high, wild 4.

Body brown above, growing gradually whiter with age, beneath and *mouth* white; *tail* white; *hair* thick, under *neck* long; teats 6, the 2 hinder spurious. Carl Linnaeus (1735)

1.1　Taxonomy and relationships with other deer

Linnaeus made an early scientific description of reindeer whilst journeying in his native Sweden. Though incorrect in the odd detail, he provided a good portrait of the species upon which this monograph is based. Present-day taxonomists consider that both Eurasian reindeer and North American caribou together form a monotypic genus with an holarctic distribution, its single representative being *Rangifer tarandus* (Linnaeus, 1758). The genus *Rangifer* belongs to the family Cervidae, the sub-order Ruminantia and the order Artiodactyla, which is one of the two groups of large herbivorous animals known as ungulates. Perissodactyls, the other group of ungulates, have one or an odd number of functional toes and simple stomachs, and include in their number equids, rhinos and tapirs. Artiodactyls, by contrast, have an even number of functional toes and ruminants, as one of the main sub-orders, have three- or four-chambered stomachs and chew the cud to complete their digestive processes. Cervids are distinguished from other ruminants such as bovids, ovids and giraffes, because most male cervids possess bony antlers which are grown and shed annually.

A major taxonomic feature of the genus *Rangifer* is that both sexes normally possess antlers (Figure 1.1). Antler shape has been used to divide the genus into two groups which transgress the continental division between reindeer and caribou (Jacobi, 1931). The group Cylindricornis have round antler beams in cross-section and occur in forest or woodland habitats, and the group Compressicornis have flattened antler beams and occur in tundra and mountain habitats. Whilst there have been other taxonomic schema for the genus (Herre, 1955; Sokolov, 1959; Flerov, 1960), Jacobi's two groups were maintained in the most recent and now accepted revision of the genus (Banfield, 1961), which lists nine extant sub-species (Table 1.1). Two of these sub-species appeared to be evolutionarily senile and became extinct early this century without the intervention of man. The present distribution of the remaining sub-species is shown in Figure 1.2, emphasising the extreme northerly latitudes that the genus occupies. In addition to the already accepted taxonomic separation between forest and tundra groups, the latter might be subdivided in future into continental and High Arctic groups. Recent taxonomic and electrophoretic evidence suggests that Peary caribou and Spitzbergen reindeer are more closely related to each other, with the likelihood of a recent common origin, than to barren-ground caribou and tundra reindeer (Røed, 1985).

Table 1.1. *Taxonomic division of the genus* Rangifer *(after Banfield, 1961)*

Latin name	Common name
Type Species:	
Rangifer tarandus	Reindeer, caribou
Group Cylindricornis:	
R. t. tarandus	Mountain or tundra reindeer
R. t. platyrhynchus	Spitzbergen reindeer
R. t. groenlandicus	Barren-ground caribou
R. t. granti	Alaskan caribou
R. t. pearyi	Peary caribou
R. t. eogroenlandicus (Extinct since *c.* 1900)	East Greenland caribou
Group Compressicornis:	
R. t. fennicus	Forest reindeer
R. t. caribou	Woodland caribou
R. t. dawsoni (Extinct since *c.* 1935)	Queen Charlotte Island caribou

Rangifer is a recent genus amongst the deer family. It originated in the early Pleistocene, centering probably in the Nearctic or in northeastern Asia, even though the earliest identifiable remains of *Rangifer* so far recovered date to 440 000 years ago and were found in Western Europe (Banfield, 1961). The genus was a prominent member of the late Pleistocene fauna which included the woolly mammoth, hairy rhinoceros, muskox, cave bear and mastodon (Zeuner, 1959). *Rangifer* survived the extinction of many of its associates and occurred either in High

Figure 1.2. Present distribution of the sub-species of *Rangifer tarandus* (after Banfield, 1961). A – *tarandus*; B – *platyrhynchus*; C – *groenlandicus*; D – *granti*; E – *pearyi*; F – *fennicus*; G – *caribou*.

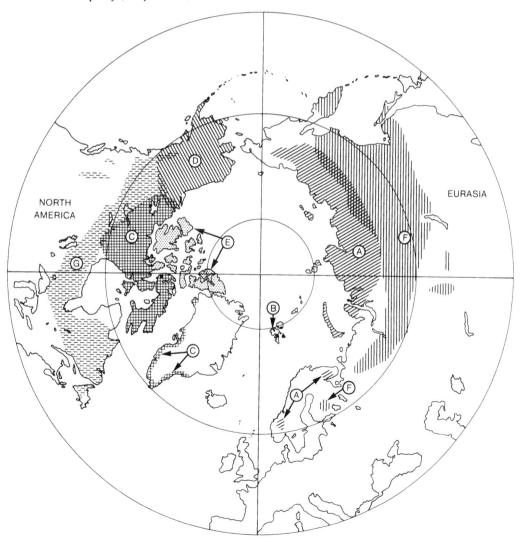

Arctic glacial refugia or at far more southerly latitudes during the Wisconsin glaciation (Macpherson, 1965). When this glaciation began to recede from its southerly limit of 40°N some 12 000 years ago, *Rangifer* surviving in more southerly ranges moved north to become important members of the present-day continental Arctic fauna. By contrast, *Rangifer* surviving in glacial refugia probably gave rise to present-day High Arctic forms, which have since colonised new areas by crossing sea-ice in winter (Banfield, 1961; Hakala, Staaland, Pulliainen & Røed, 1985).

Muskox are amongst the contemporary species of ungulate which survived the Pleistocene extinction and with which *Rangifer* overlap in their present-day range. Muskox prefer the oasis areas within the High Arctic desert and now only occur naturally in North America and Greenland (Tener, 1965), having been overhunted and reintroduced into various areas of Eurasia. Mountain sheep of several species and races occur in North America and Eurasia but are very restricted and occur in areas of high altitude dominated by grasses and sedges (Geist, 1971), and saiga are caprids which have survived in the steppe regions of Eurasia at the southern end of *Rangifer*'s range (Bannikov, Zhirmov, Lebedeva & Fandeev, 1967). Of these present-day arctic ungulates, *Rangifer* shows the greatest niche breadth, and occurs in most habitats of the arctic-

Figure 1.3. Relationship between shoulder height and antler length of males in different species of deer (after Clutton-Brock, Albon & Harvey, 1980). Squares (A), small (≤2) breeding groups; circles (B), medium (3–5) groups; triangles (C), large (≥6) groups, including *Rangifer*.

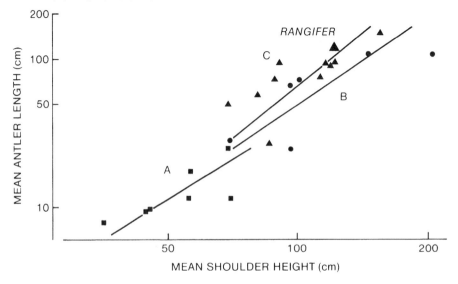

alpine biome, excepting steeply graded mountain slopes and cliffs (White *et al.*, 1981).

Also of interest to this study is the relationship between the ecology of *Rangifer* and other contemporary cervids. The various species of tropical and temperate deer have been classified according to the sizes of their breeding groups, and *Rangifer* is amongst those species having large groups (Clutton-Brock *et al.*, 1980, 1982). Characteristics that favour polygyny include large body size of males relative to females and large size of antlers in males. These characteristics are more enhanced in species living in large groups, as shown for *Rangifer* in Figure 1.3. Amongst temperate and Arctic species only medium and large breeding group sizes are represented (Table 1.2). *Alces alces* is the next most northerly species, which also belongs to a monotypic genus that is holarctic in its distribution. It lives in cold northern forests but some confusion exists because it is referred to as 'moose' in North America and 'elk' in Europe. Members of the genus *Cervus* occur in both North America and Eurasia, and are temperate and tropical in their distribution. However, only two of the nine species are of major interest in temperate areas: red deer (*Cervus elaphus*) and wapiti (*C. elaphus canadensis*), which is usually referred to as the 'elk' in North America. Hereafter, I will refer to *A. alces* as moose and *C. e.canadensis* as wapiti, in the hope of avoiding further confusion. Single genera occur in each of North America and Eurasia and their only representatives are white-tailed deer, *Odocoileus virginianus*, mule or black-tailed deer, *O. hemionus*, and roe deer, *Capreolus capreolus*, which all inhabit various niches in temperate woodlands. A final monotypic genus, the fallow deer (*Dama dama*), originated in the Mediterranean area, but is often included amongst temperate species, having been introduced widely to forest and open areas of Eurasia and North America. The marked degree of polygyny seen amongst reindeer and caribou will be discussed in chapters of Part II in relation to the other species of temperate cervid.

1.2   Habitats and ecology

The various sub-species of *Rangifer* occur at latitudes varying from 45° to 80°N (Figure 1.2) and in habitats ranging from taiga woodlands to High Arctic islands. At these latitudes, snow and cold are dominant factors in the lives of terrestrial animals for much of the year (Pruitt, 1960b; Formosov, 1964). In adapting to them, *Rangifer* has evolved many interesting morphological characteristics (Skjenneberg & Slagsvold, 1968; Kelsall, 1968; Baskin, 1970; Banfield, 1977). Their

Table 1.2. *Arctic and temperate species of contemporary cervids, showing their breeding group size (B, 3–5; C, ≥6), and habitat type (from Clutton-Brock, Guinness & Albon, 1982), and origin (Holarctic, Nearctic, and Palearctic)*

Species	Breeding group size	Open country	Broken canopy forests, woodland or fringe	Closed-canopy forests	Origin
Rangifer tarandus (reindeer and caribou)	C	—	—		H
Cervus elaphus canadensis (wapiti)	C	—	—		N
Cervus elaphus (red deer)	C	—	—		P
Dama dama (fallow deer)	C	—	—		P
Odocoileus virginianus (white-tailed deer)	B		—		N
Odocoileus hemionus (mule deer, black-tailed deer)	B		—		N
Alces alces (moose)	B		—	—	H
Capreolus capreolus (roe deer)	B		—	—	P

winter coat is long and dense, and comprised of air-filled guard hairs and a close underfur for extra insulation. Their crescentic hooves are unusually large and facilitate both travel across snow and boggy ground, and also digging through the snow for winter forage. Their legs are long and powerful enabling the species to cover large distances and also to swim strongly.

Woodland sub-species are relatively sedentary and have short altitudinal migrations (Edwards & Ritcey, 1960). In the High Arctic, Spitzbergen reindeer now have a restricted range on Svalbard (Hjeljord, 1973; Tyler, 1987a), but Peary caribou frequently travel between islands on sea-ice (Miller, Russell & Gunn, 1977). Migrations are most extensive in continental tundra sub-species (Geller & Borzhanov, 1984), and especially in barren-ground caribou which cover remarkable distances of up to 1000 km on each migration in moving to the tundra in the spring and returning to taiga woodland in autumn (Kelsall, 1968). Calves are born in the tundra during, or at the end of, the spring migration in May or June. They are extremely precocious and capable of travel within a few hours of birth and are the only species of cervid not to have a spotted coat at birth.

Migrations, whether short or extensive, are necessary to exploit seasonal differences in availability of forage. Tundra vegetation is highly nutritious in its short summer growing season and fly harassment is reduced in open, windswept areas (Klein, 1970a; White *et al.*, 1975). During winter, when forage is unavailable over much of the tundra, reindeer and caribou have to find access to food that is beneath snow shallow enough for them to dig through. *Rangifer* is able to live over much of the north because it has adapted to a winter diet in which lichens (Figure 1.4) usually predominate, and there is little competition with other Arctic herbivores for this niche (Klein, 1980a). As lichens have slow growth rates and recover their former abundance several decades after extensive damage due to grazing or trampling (Palmer & Rouse, 1945; Pegau, 1970a), there is a delicate interaction between *Rangifer* and its habitat. This has led several authors (Leopold & Darling, 1953; Andreev, 1977) and many traditionalists to assume that lichens play a central role in limiting numbers of reindeer and caribou in many areas of the holarctic. The underlying mechanism of population regulation presumably lies in long-term cycles of lichen abundance, overutilisation and recovery.

The climate of Arctic regions also shows long-term fluctuations, and *Rangifer* generally shows marked and cyclical fluctuations in its numbers

Figure 1.4. 'Reindeer' lichens of the genus *Cladonia* in conifer woodlands in Norway. Reindeer and caribou are the only northern ungulate to occupy the lichen-based food niche in the Arctic.

(Elton, 1942; Vibe, 1967). Unfortunately accurate long-term records of numbers only extend back to the 1950s, but historical evidence certainly suggests that long-term fluctuations do occur (Meldgaard, 1986). In the High Arctic, extinction of East Greenland caribou in the early 1900s was probably associated with a climatic change that led to the absence of sea-ice and heavy snow accumulation in winter (Vibe, 1967). Furthermore, barren-ground caribou in West Greenland declined at this time but recovered to reach peak numbers in the 1960s, before experiencing another rapid decline (Strandgaard, Holthe, Lassen & Thing, 1983). Peary caribou are also experiencing a decline in numbers (Gunn, Miller & Thomas, 1981). Tundra reindeer on the Novosibirsky Islands experienced a decline in the 1920s because failure of the sea-ice prevented their traditional migration to the mainland but they have since recovered in number (Kishchinskii, 1971, 1984; Klein & Kuzyakin, 1982). Long-term changes in numbers have also been deduced for more southerly populations (Edwards, 1956; Skoog, 1968; Kelsall, 1968; Hemming, 1975), but are often obscured by the activities of man (Section 1.3). However, the limited evidence available is sufficient to infer that long-term cycles in both climate and food availability have had a major influence on the ecology of *Rangifer* in many areas (White *et al.*, 1981; Meldgaard, 1986).

Reindeer and caribou historically have shared much of their range with predators, the exception being on certain High Arctic islands like Svalbard. Apart from man, the most notable predators in continental areas are the various sub-species of timber and tundra wolves, which have an holarctic distribution and have co-evolved with *Rangifer* (Figure 1.5) and other cervids throughout their range (Mech, 1970). In areas where wolves have not been eliminated, reindeer and caribou are an important part of their diet (Murie, 1944; Kelsall, 1968; Zhigunov, 1969; Parker & Luttich, 1986). Predators such as lynx and wolverine are locally important (Skjenneberg & Slagsvold, 1968; Bergerud, 1971; Semenov-Tian-Shanskii, 1975) while others like bears and golden eagles are of little importance (Kelsall, 1968). Predation by wolves and other species is frequently directed at immature members of their prey species (Pimlott, 1967; Bergerud, 1971; Miller & Broughton, 1974; Baskin, 1983). This has fired a strong debate amongst biologists as to whether predation constitutes the major limiting factor for the size of northern prey populations. For example, all suggestions that overutilisation or destruction of habitat and lichens limits numbers of caribou in continental North America have been firmly rejected by Bergerud (1974c, 1980a, 1983). Studies of introduced reindeer on islands are of interest in this debate. Reindeer on South Georgia and other islands have acted as control experiments that document population responses in the absence of predators. Further interest derives from comparisons offered with certain other populations of *Rangifer*. These include island and High Arctic populations that naturally lack or have few predators and clearly are limited by other factors, and those populations from which predators have been eliminated by man.

1.3   Relationships with man

The association of man with reindeer and caribou has been referred to already, and together with lichens, snow and predators, he now exerts a dominant influence upon their ecology. Ancient reindeer remains, found in most countries of Western Europe and as far south as Spain and Italy, have almost invariably been associated with human cultures (Charlesworth, 1957). Cave remains have shown that upper Paleolithic peoples depended heavily upon reindeer products, and the

Figure 1.5. A wolf eating a caribou carcass in Alaska. Wolves are important natural predators of *Rangifer* in continental areas. (Photograph by D.D. Roby.)

species has appeared prominently in cave art and rock carvings in many localities (Sieveking, 1979; Figure 1.6). Seasonal migrations have been a feature of the biology of *Rangifer* since at least Paleolithic times, for it is clear that the cave dwellers of the Pyrenees adopted a strategy of herd-following, migrating with the reindeer to either Atlantic or Mediterranean seaboards in summer (Bahn, 1977). This early tradition of herd-following culminated in the semi-domestication of reindeer, a way of life that has been practised by northern peoples such as Lapps, Samoyeds, Yakagir and Tungus throughout Eurasia for the past 2000 years or so (Lindgren, 1935; Utsi, 1948; Zeuner, 1963; Ingold, 1980; Syroechkovskii, 1984; Baskin, 1986).

Herding practices throughout Eurasia have been of great importance, for distinctions between wild sub-species have largely been broken down by breeders of herded reindeer (Figure 1.7). Also as a result of conflicts between reindeer-breeding and wild reindeer, the remaining populations of the wild sub-species and their predators have become reduced and isolated into limited areas (Klein, 1980b; Skrobov, 1984). For example, in areas of Lappish influence, wild mountain reindeer are now only found in the mountains of southern Norway (and these are probably derived from herded animals which escaped: Krafft, 1981) and on the Kola Peninsula of USSR (Semenov-Tian-Shanskii, 1975); wild forest reindeer are similarly limited to a small area of Karelia lying between USSR and Finland (Nieminen, 1980a). Herded animals have now replaced wild reindeer throughout most of the suitable range in Fennoscandavia and USSR, which possesses by far the largest populations of both wild and herded reindeer (Table 1.3) of any single country (Syroechkovskii, 1984; Razmakhnin, 1986).

Figure 1.6. Stone/Bronze Age rock-carving of reindeer from Sagelva, Norway, 2000 to 500 B.C. (from Hagen, 1965).

Caribou have not been domesticated, but various groups of North American Indians and Inuit also have depended upon them for a long time (Spiess, 1979). Inuit led a sedentary life in areas where caribou were present all year round, but many Indians followed migrations: all made their kills by hunting (Burch, 1975; Sharp, 1977). Before the advent of modern firearms, the techniques employed included spearing swimming caribou from canoes and kayaks, trapping in pitfalls and drift fences and stalking or ambushing of moving animals (Kelsall, 1968). Upon settlement of North America by white man, and the subsequent introduction of firearms to native peoples, caribou have generally shown major declines in number through most of their ranges, due in large part to overhunting (Elton, 1942; Skoog, 1968; Kelsall, 1968; Bergerud, 1974c; Miller, 1983). This precipitated the 'caribou crisis', and even today a number of mainland herds are decreasing in size (Miller, 1983; Williams & Heard, 1986). Even island populations in the High Arctic have been affected by the predations of man. Spitzbergen reindeer were nearly exterminated in the early 1900s (Hjeljord, 1975), numbers of tundra reindeer on Russian islands were seriously affected by indiscriminate hunting (Parovshchikov, 1965; Klein & Kuzyakin, 1982), and a combin-

Figure 1.7. Kautokeino Lapps and their reindeer in a corrall in northern Norway. Breeders of semi-domestic reindeer have broken down distinctions between sub-species and coat colours show many variations.

Table 1.3. *Status of Rangifer tarandus in its natural range (based mainly on Williams & Heard, 1986)*

| Country | Wild reindeer and caribou | | | Numbers of herded reindeer | Additional sources |
	Sub-species	Numbers	Status		
Norway	R. t. tarandus	42000	Stable	200000	Reimers et al. (1980)
	R. t. platyrhynchus	8800	Stable		
Sweden				200000	Nordkvist (1980)
Finland	R. t. fennicus	9000[a]	Stable	150000	Pulliainen (1980); Pulliainen & Siivonen (1980); Danilov & Markovsky (1983)
USSR	R. t. tarandus	800000	Increasing	>2.5m	Semenov-Tian-Shanskii (1975); Borzjonov et al. (1979); Klein & Kuzyakin (1982)
	R. t. fennicus	180000	? Declining		
China				1000	Ma Yi-ching (1983)
Alaska	R. t. granti	500000[a]	Increasing		Davis (1980)
Canada	R. t. groenlandicus	1.3m[a]	Increasing		Bergerud (1980b); Calef (1980); Gunn et al. (1981)
	R. t. caribou	750000[a]	Increasing		
	R. t. pearyi	25000	Declining		
Greenland	R. t. groenlandicus	15000	Declining		Strandgaard et al. (1983)

[a] Includes numbers of that sub-species occurring in a contiguous population in an adjacent country.

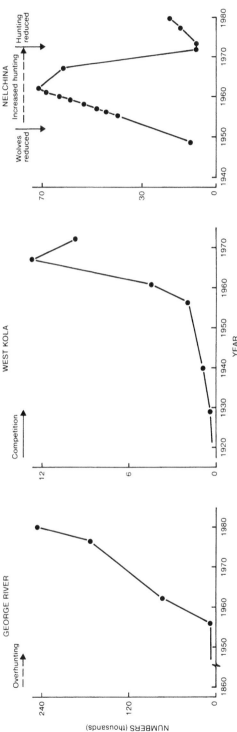

Figure 1.8. Examples of irruptions in established herds of woodland caribou, tundra reindeer and Alaskan caribou. The George River herd was overhunted for many years; the West Kola herd was subjected to competition from herded reindeer until protected in 1929; wolves were greatly reduced on the Nelchina range in the early 1950s, whilst hunter harvests increased to reach a peak in the early 1970s (data from: Elton, 1942; Parker 1981; Semenov-Tian-Shanskii, 1975; Hemming, 1975; Bergerud, 1980a; Van Ballenberghe, 1985).

ation of heavy winters and overhunting resulted in extirpation of caribou in NW Greenland (Roby, Thing & Brink, 1984).

In contrast, man has attempted to eliminate predators from many reindeer and caribou ranges to protect his resource. All other things being equal, predator removal (including cessation of heavy hunting by man) usually results in a dramatic increase of numbers of ungulates (Rasmussen, 1941; Leopold, 1943), known as an irruption. ('Irruption' has also been spelled 'eruption' and both words used in the same context. However, only American dictionaries such as Webster's New Collegiate Dictionary have a specialised definition for 'irruptions in animal

Figure 1.9. An irruption in Spitzbergen reindeer after hunting was banned in 1925. As numbers of reindeer increased, they dispersed to new areas shown by arrows: 1 and 2 in 1950s; 3 in 1970s, 4 after 1975 (from Tyler, 1987a).

populations'. Hereafter, and with misgivings arising from implications of its derivative prefix, 'irruption' will be used and other authors' usage of 'eruption' converted to this spelling.) It is probably true to say that most irruptions which now occur amongst established populations of ungulates arise largely as a result of human interference. Ungulates may be affected directly, or indirectly by actions that affect the ecosystem; the interference may be a current condition or an historic one that still affects the population (Peek, 1980).

Even though long-term fluctuations might have occurred in continental reindeer and caribou populations (Kelsall, 1968; Skoog, 1968; Hemming, 1975; Haber & Walters, 1980), these would be hard to disentangle from fluctuations recorded since the first accurate censuses were instigated, for many of them are man-induced (Bergerud, Jakimchuk & Carruthers, 1984; Syroechkovskii, 1984). A few examples of such irruptions from amongst established populations of tundra and Spitzbergen reindeer and of woodland and Alaskan caribou are shown in Figures 1.8 and 1.9. From these and the foregoing discussion it is apparent that man now holds the key to the status of most populations of reindeer and caribou.

1.4   Summary

1. Reindeer and caribou are a single species with an holarctic distribution. They live at high latitudes and in large groups compared with other species of deer.
2. Reindeer and caribou usually undergo annual migrations, and depend in winter upon lichens which they dig for beneath the snow. Predators and especially wolves have co-evolved with them on much of their range.
3. Reindeer and caribou populations have undergone long-term fluctuations in number. Man has herded reindeer and hunted caribou for several centuries, and now exerts a dominant influence upon their ecology.

2

The introduction of mammals to new environments

There are thousands and thousands of small, remote islands that have been too far from the busy evolutionary centres of the continents to acquire more than a sprinkling of accidental immigrants before the arrival of man began to make this process of dispersal so much easier and faster.
Charles Elton (1958)

2.1   The irruption of introduced ungulates

Elton (1958) made an early study of the introduction of species to new habitats. Ungulates are amongst the groups of mammal that have been introduced widely throughout the world (Lever, 1985). Many introductions fail at source, so that the introduced species does not become established in its new habitat (de Vos, Manville & Van Gelder, 1956). However, if a new habitat is suitable for their requirements, an exotic population immediately faces a situation of great instability upon introduction. There is normally an abundant food supply, limited competition with native species and virtual freedom from predation (Elton, 1958). The usual response in such a situation is an irruption, which follows a well-defined sequence of stages that was first defined in New Zealand (Riney, 1964). This island group once lacked terrestrial mammals, but has since had many introductions made to it: 53 different species of mammal have at one time or another been liberated, and at least 34, including 11 species of ungulate, have since become established (Wodzicki, 1950, 1961; Caughley, 1976a). Riney proposed that introduced ungulates normally adjust to their new habitat with a single irruptive oscillation comprised of a sequence of four arbitrary stages, and which does not differ in principle from fluctuations seen in established populations (Riney, 1964). Habitat changes occurring in New Zealand after the introduction of herbivores were classified into a sequence also of four stages (Howard, 1964). The combination of Riney's and Howard's suggestions therefore result in a sequence where animals and vegetation (and thus the carrying capacity of the habitat) follow reciprocal trajectories (Figure 2.1).

During stage 1, there is a progressive increase in numbers in response to the discrepancy between the carrying capacity of the habitat and the numbers of ungulates present. The death rate is low and as new generations reproduce, the population curve increases rapidly. Eventually preferred species in plant communities become overutilised. During stage 2, the population exceeds its carrying capacity as more extensive areas of vegetation are overutilised. Physical condition at critical seasons drops noticeably; as juvenile death rate increases the population starts to level off in number. In stage 3, the population begins to decline, especially when an element of the environment such as climate becomes critical. Latterly, some plant species less susceptible to browsing or grazing pressure may begin to recover as the animal density declines to a level more nearly compatible with the current carrying capacity of the habitat. Stage 4 commences when a degree of stability has developed between numbers of ungulates and the new carrying capacity of the habitat. Population density at this stage remains lower than peak density, because plant communities have a lower cover of preferred species than when the population was first introduced.

Once this complete adjustment has taken place, the introduced

Figure 2.1. Diagram of the stages occurring during an irruptive oscillation after introduction to a new area, extending into a phase of post-decline stability (after Riney, 1964; Howard, 1964). CC = carrying capacity; 1–4 = stages of the oscillation which are explained in the text.

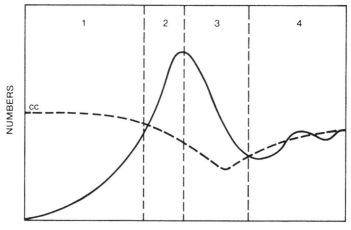

TIME AFTER INTRODUCTION

population, ecologically, is the same as any established population. Not until an exotic species moves beyond the initial phase can the success or otherwise of the introduction be permanently evaluated. Further irruptions of a dampened nature may occur only after creation of another discrepancy between carrying capacity and numbers present. They might result from an increase in carrying capacity (from, for example, removal of intraspecific competitors or habitat improvement) or a decrease in population size (following a serious disease or cessation of heavy hunting), exactly as discussed earlier for established populations. Even though in stage 4 the number of introduced ungulates tends to fluctuate from year to year, there is usually little chance of the stage 2 situation being repeated because the composition of the vegetation may have altered as a result of selective utilisation.

This basic model of an irruptive oscillation was developed further (Riney, 1964). On islands the size of New Zealand's, ungulates can disperse away from their point of liberation, probably in response to an actual or impending shortage of food (Lack, 1954). It was thus suggested that the sequence of events in Figure 2.1 occurred successively in each new area occupied (Figure 2.2). At the newest dispersal front, and furthest from the point of liberation, density starts to increase (stage 1); further back in the range density is at its peak (stage 2); nearest the point of liberation the population attains relative stability at a lower density (stage 4). The sequence of stages following liberation could thus be observed either spatially or temporally, at one place over a period of time, or at one time over a range of distance.

Figure 2.2. The left graph is the irruptive oscillation in Figure 2.1. Graphs to the right represent the occurrence of the same sequence in a similar habitat, but at later dates as the population disperses further from the point of liberation (after Riney, 1964; Caughley, 1970).

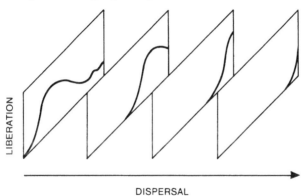

LIBERATION

DISPERSAL

To test the irruptive model, Himalayan thar (also spelt tahr), a goat-like bovid that had been introduced to the South Island of New Zealand (Figure 2.3), were studied by Caughley (1970). To make the study independent of assumptions, the stages of the model were redefined, to exclude vegetational changes and to use only attributes of the

Figure 2.3. Himalayan thar were introduced to the South Island of New Zealand in 1904. Their irruption has been studied extensively. (Photograph by courtesy of the New Zealand Forest Service.)

population. The four stages thus became: (1) initial increase, until attainment of the initial peak; (2) initial stabilisation, until the population commences a significant decrease in number; (3) decline; and (4) post-decline. If the model was to be proved correct, the following had to be demonstrated: rate of population increase had to be positive at stage 1, zero or close to zero at stages 2 and 4, and negative at stage 3; survival and birth rates, and fat reserves had to be highest at stage 1, lowest at stage 3, and intermediate at stages 2 and 4.

These population attributes were sampled and rates of increase were calculated within four predetermined areas at progressively greater distances from the original point of liberation of thar (Caughley, 1970). Thar numbers were known to be increasing at the dispersal front, and to have declined near the point of liberation: the critical test of the model lay in confirming the proposed stage 2 and 3 areas differed. Such differences were demonstrated, but though birth rates varied at each stage, the major influence on the rate of increase was variations of the death rate, especially in young thar (Caughley, 1970). The food supply available to thar also was shown to decrease along the transect from dispersal front to the point of liberation. This strongly supported suggestions that increased availability of food initiated the upswing of an irruptive oscillation in ungulates, and that such an upswing was terminated by overgrazing or overbrowsing (Howard, 1964; Riney, 1964).

The thoroughness of Caughley's study, and his later development of mathematical models of vegetation–herbivore interactions (Caughley, 1976a, 1977), might appear to have left few topics in this field for others to confirm or investigate. However, several additional problems are either highlighted in his review of other ungulate irruptive oscillations or not considered in his study, as the following two major examples will serve to illustrate. First, on continents or large islands, such as New Zealand, introduced ungulates have available to them a considerable area for movement away from the point of liberation. In this situation, Figure 2.2 is a realistic depiction of events, and the method of sampling at one time over a range of distance is quite valid (Caughley, 1970). The situation on small islands, in which the areas available are small compared with an ungulate's ability to travel, is quite likely to be different. Secondly, most introductions which did not fail initially (de Vos *et al.*, 1956) and which entered an irruptive oscillation have followed Riney's and Howard's model (Caughley, 1970). Two notable exceptions have been the introductions of reindeer made to two small islands in the Bering Sea, where crash die-offs resulted on St Paul and St Matthew islands (Scheffer, 1951; Klein, 1968). However, reindeer introductions elsewhere have not ended with quite such dramatic results. Thus reindeer introduced to small islands provide an unequalled opportunity to study a range of population responses made under different ecological conditions.

2.2   Introductions of reindeer

Reindeer have been introduced to new habitats in the Arctic and sub-Arctic (Figure 2.4), as a basis on which to form a new industry for native peoples. Unsuccessful introductions, which never appear to have entered an irruptive oscillation, were made to the Slave River area,

Figure 2.4. Continental areas and islands in the Arctic to which reindeer have been introduced (data from Table 2.1).

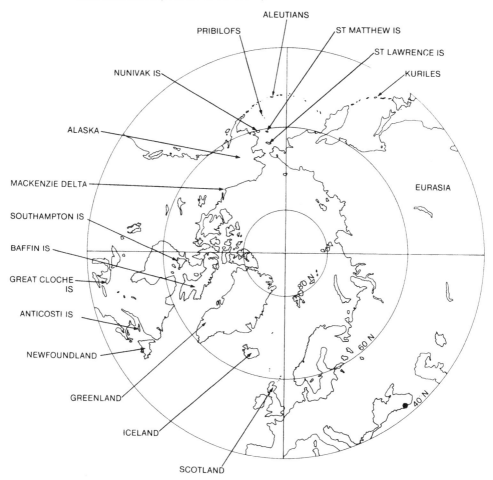

North West Territories, Canada in 1911 and to Michigan, USA in 1922
(Lindgren & Utsi, 1948). Two other introductions that have been made
to continental areas have been successful. From 1891 to 1902, nearly 1300
reindeer were brought to Alaska from Siberia to replace the formerly
abundant caribou upon which the subsistence hunters depended and
which had declined seriously in the late 1800s (Skoog, 1968). The
reindeer increased in number dramatically, spreading out from the areas
of introduction and reaching maximum numbers of over half a million in
1930–8 (Figure 2.5). By the 1950s numbers had declined to only about
25 000, and the range occupied contracted markedly so that a viable
industry is now based on about 20–30 000 head of reindeer which occur in
20 herds mainly occupying the Seward Peninsula (Davis, 1980; Naylor,
Stern, Thomas & Arobio, 1980).

After earlier attempts to introduce reindeer to mainland Canada, 2370
reindeer were driven to the Mackenzie Delta area in 1935 from Alaska
over a trail that had taken six years to negotiate (Treude, 1968). By
1942–3 there were over 9000 reindeer, after which numbers fluctuated

Figure 2.5. Population change and range occupancy of reindeer introduced to
mainland Alaska (data from Leopold & Darling, 1953; Davis, 1980).

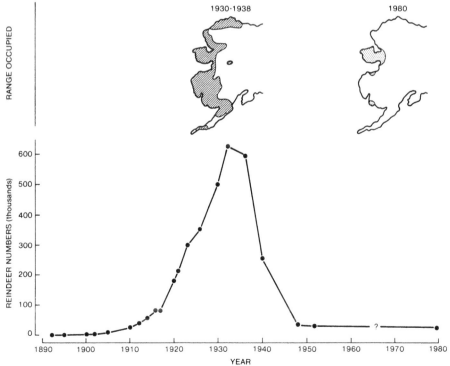

Table 2.1. Results of introductions of reindeer and two of caribou[a] to islands throughout the world

Island	(Country/area)	Latitude	Size	Date of introduction	Number introduced	Entered an irruptive oscillation	Successful	Source
Aleutians								
Adak[a]	(Bering Sea)	52°N	Small	1958/9	24	Yes	Yes	Jones (1966); Davis (1980); Klein (personal communication)
Atka		52°N	Small	1913/14	40	Yes	Yes	Grenfell (1934); Treude (1968)
Unmak		54°N	Large	1913/14	36	Yes	Yes	Treude (1968)
Anticosti	(Canada)	49°N	Large	1918/19	145	Yes	No	Grenfell (1934); Treude (1968)
Baffin	(Canada)	65°N	Large	1921	550	No	No	Treude (1968)
Great Cloche	(Canada)	46°N	Small	1969	14	No	No	Anderson (1971)
Greenland		61–77°N	'Small'	1952–73	263	Yes	Yes	Thing (1980); Strandgaard et al. (1983)
Iceland		63–65°N	Large	1771	13–14	No	No	Thórisson (1983)
				1777	30	Yes	No	
				1784	35	Yes	No	
				1787	35	Yes	Yes	
Kerguelen	(Subantarctic)	49°S	Large	1955	7	Yes	Probably	Lesel & Derenne (1975); Pascal (1982)
Kuriles	(Japan/USSR)							
Shinshiru		46°N	Small	1924	10	Yes	Uncertain	Whitehead (1972)
Navarino	(Chile)	53°S	Large	1944	?	No	No	Grenfell (1934)
Newfoundland	(Canada)	51°N	Large	1908	300	Yes	No	Hanson (1952); Davis (1980)
Nunivak	(Bering Sea)	60°N	Large	1920	99	Yes	Yes	Scheffer (1951)
Pribilofs								
St Paul	(Bering Sea)	57°N	Small	1911	25	Yes	Yes	Hadwen & Palmer (1922); Davis (1980)
St George		56°N	Small	1911	15	Yes	No	Klein (1968)
St Lawrence	(Bering Sea)	63°N	Large	1900	76	Yes	Yes	Lindgren & Utsi (1980)
St Matthew	(Bering Sea)	60°N	Small	1944	29	Yes	No	Parker (1975); Williams & Heard (1986)
Scotland	(UK)	57°N	Large	1952–5	35	Yes	Yes	This study
Southampton[a]	(Canada)	64°N	Large	1967	48	Yes	Probably	
South Georgia	(Subantarctic)	54°S	'Small'	1911	10	Yes	Yes	
				1911	5	No	No	
				1925	7	Yes	Yes	

between 5000 and 8500 until 1966, when they decreased to 2400 (Krebs, 1961; Scotter, 1972). Since then the herd has changed ownership and was run by a government agency before being sold into private hands, and during this period reindeer numbers have increased to 15000 (Stager, 1984). Reindeer from both Canadian and Alaskan introductions have been slaughtered and subjected to a variety of herding practices and policy decisions over the years (Treude, 1968; Thomas & Arobio, 1983). The Canadian introduction initially appeared to enter an irruption and then declined, but has now entered a second and larger irruption, following greatly improved management practices (Stager, 1984). However, the Alaskan introduction has shown a more characteristic irruptive oscillation (Figure 2.5).

The spread of reindeer away from the point of liberation in continental areas (Figure 2.5) and its likelihood on large islands (cf. thar in New Zealand) has led me to classify islands to which reindeer and caribou have been introduced (Table 2.1) as either large or small. However, the distinction is hard to draw, both absolutely in terms of island size, and relatively due to the presence of physical barriers. Several large islands in polar regions such as Greenland are subdivided into 'small islands' by glaciers. Several of the island introductions clearly failed to enter an irruptive oscillation, and others have failed in the long term (Table 2.1), supporting contentions that many mammalian introductions fail and that the success of an introduction cannot be assessed until after the initial irruptive oscillation (de Vos *et al.*, 1956; Riney, 1964). Large islands such as Iceland, Nunivak and St Lawrence have all supported successful introductions which went through irruptive oscillations but still have sufficient reindeer to support native industries or hunting (Figure 2.6; see also Figure 8.4). The most dramatic results stem from introductions of reindeer made to the Pribilof Islands and to St Matthew Island (Scheffer, 1951; Klein, 1968). All of these are classified as small islands, where there was probably a single irruptive oscillation in a relatively restricted area (Figure 2.6).

On St Paul and St George islands, an annual inventory of reindeer numbers was made from the date of their introduction in 1911 until 1950 (Scheffer, 1951). On St Paul (107 km^2) the original population of 25 reindeer increased in number steadily to reach 2000 animals in 1938 but thereafter declined, and by 1950 there were only eight reindeer. This small population has since increased and stabilised at around 250 animals (Davis, 1980). On St George (91 km^2) the original population of 15 reindeer increased to reach 220 animals by 1922 (a similar contemporaneous total to that on St Paul), but thereafter decreased and fluctuated between 77 and 10 reindeer from 1925 to 1950; no recent data for St George are available (Davis, 1980). No definite reason could be advanced for the differences between St Paul and St George islands. It was suggested that some ecological difference caused the latter population to undergo a more dampened fluctuation, even though both islands are relatively close neighbours and of a similar size (Scheffer, 1951).

On St Matthew Island (332 km^2), the original population of 29

Figure 2.6. Population changes of reindeer introduced to various northern islands. An annual inventory was carried out on St Paul (solid line) and St George (dashed line) islands but only a few data points are shown for reasons of clarity (data from: Scheffer, 1951; Klein, 1968; Davis, 1980; D.R. Klein, personal communication).

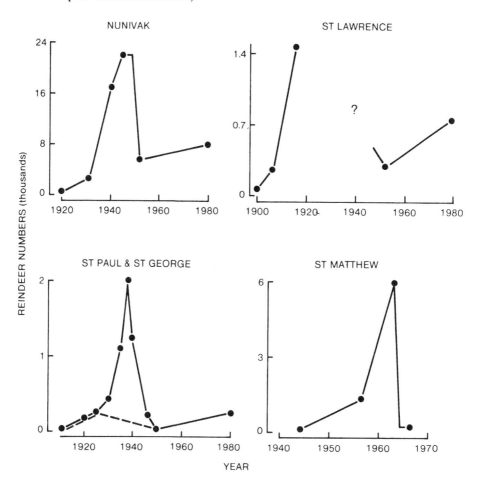

reindeer introduced in 1944 increased very rapidly, reaching 1350 animals in 1957 and 6000 by 1963 (Klein, 1968). However, there was then a crash die-off and in 1966 there were only 42 reindeer on the island. Both these die-offs were attributed to an interaction of two main factors (Scheffer, 1951; Klein, 1968). First, reindeer depended on lichens as their winter forage, and these had become severely overgrazed as reindeer numbers built up; secondly, the die-offs occurred during periods when winters were particularly severe. Additionally, reindeer were in relatively poor condition going into winter due to competition for high quality summer forage. Weather conditions clearly were not the sole factor responsible for the die-offs, as other island populations of reindeer in the Bering Sea were not similarly affected (Klein, 1968).

Therefore, as with other populations of introduced ungulates, it may be concluded that the upswing of the irruptive oscillation of reindeer on St Paul and St Matthew islands was terminated by overgrazing (cf. Caughley, 1970). However, in contrast to most other irruptive oscillations, these populations entered a limit cycle to extinction (Caughley, 1970, 1981). They showed an exaggeration of stage 3 and omitted stage 4 from their oscillations. A transition from stage 3 to stage 4 can occur only if an equilibrium between continued availability of forage on an overgrazed range at critical seasons and lower density of animals on that area is soon reached. The intrinsic ability of vegetation to respond favourably to heavy utilisation and to recover some of its former abundance are of vital importance in this respect. Any such recovery would proceed more quickly if utilisation pressure is reduced additionally by emigration of some of the population to a new area (as in Figure 2.5). However, even when reindeer are absent, lichens recover very slowly from heavy grazing and trampling (Palmer & Rouse, 1945; Hustich, 1951). Their slow growth and recovery time, the apparent lack of availability of other suitable winter forage and the obvious lack of opportunity for emigration would appear to have been reasons why reindeer on St Paul and St Matthew islands did not quickly achieve the balance necessary to attain stage 4. Whilst firm data are lacking, a further look at Table 2.1 shows that other introductions failed after entering an irruptive oscillation and probably followed a similar course to those described by Scheffer (1951) and Klein (1968). Apart from the small sample of reindeer collected by Klein (1968), there has been no systematic study of demographic characteristics of island populations of introduced reindeer in the Arctic. It is against this background that introductions made to subantarctic islands are of interest.

2.3   **Introductions to southern islands**

The islands of the Southern Ocean lack indigenous terrestrial mammals. However, species varying in size from mice to horses and cattle have been introduced to them since the 19th century, and several have now become established on different subantarctic and southern temperate islands (Figure 2.7). The ecological impact of successfully

Figure 2.7. The Antarctic continent and surrounding Southern Ocean, showing the introduced species of mammals which have been established on subantarctic and southern temperate islands (data from: Holdgate & Wace, 1961; Bonner, 1984a; Clark & Dingwall, 1985; Leader-Williams, 1985).

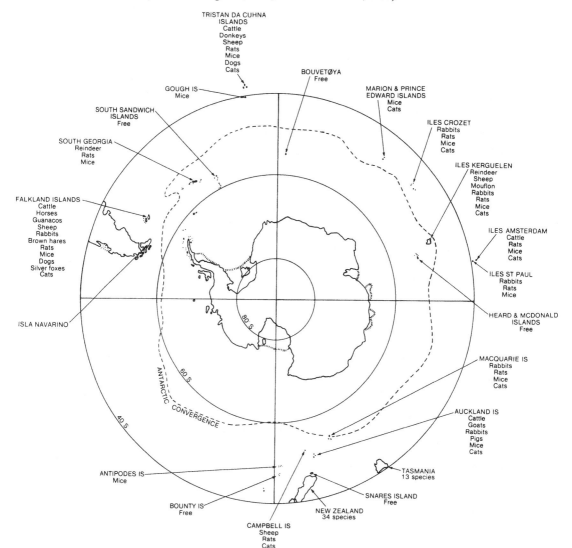

introduced mammals varies enormously. It is not always possible to predict the consequences of an introduction, because an exotic species rarely occupies the niche which the introducer expected it to fill (de Vos *et al.*, 1956), if indeed this was thought about at all! This is certainly true of several of the introductions made to southern islands, which have resulted in noticeable effects on indigenous floras and faunas, and caused concern to conservationists (Holdgate & Wace, 1961; Bonner, 1984a). Thus cats, introduced to control rodents or as companion animals for men living on remote stations, have run wild and prey heavily upon, and have sometimes eliminated, certain species amongst native avifaunas (Jones, 1977). Herbivores such as rabbits, sheep and goats have severely damaged areas of vegetation (Costin & Moore, 1960; Wilson & Orwin, 1964; Campbell & Rudge, 1984).

An advantage to ecologists of these introductions lies in the climatic severity and relative simplicity of terrestrial ecosystems on southern islands. Therefore, these introductions are an extreme test of adaptation. Some superb natural experiments have been created, and many still await fuller investigations. Amongst these are the introduction of reindeer that were made to Iles Kerguelen in 1955. According to the only reports available, the population on the main island of Grande Terre is still increasing (Lesel, 1967; Lesel & Derenne, 1975; Pascal, 1982). One other introduction to another southern hemisphere island, Isla Navarino, made in 1944 has failed to become established. The following pages will concentrate upon the reindeer of South Georgia.

2.4 Summary

1. Upon introduction to any new habitat lacking competitors and predators, an ungulate population faces a situation of instability, arising from an abundant food supply. The usual response in such a situation is an irruption.
2. Introduced reindeer provide comparative opportunities to study the irruptions of ungulates, having been liberated into many new habitats.
3. Introduced mammals are widespread on islands in the Southern Ocean, and have seriously affected native faunas and floras.

3

South Georgia: the island and its reindeer

The Wild rocks raised their lofty summits till they were lost in the Clouds and the Vallies laid buried in everlasting snow. Not a tree or shrub was to be seen, no not even big enough to make a toothpick.

Captain James Cook on discovering South Georgia in 1775 (Beaglehole, 1961)

3.1   Environment and natural history

The subantarctic island of South Georgia is situated some 2000 km east of Tierra del Fuego in the South Atlantic Ocean at latitude 54°–55°S, longitude 35°–38°W. It is the largest island of the Scotia Ridge, and is about 160 km long and varying in width from 5 to 40 km, with its axis trending northwest to southeast (Figures 2.6, 3.1). Geologically, most of the island consists of sedimentary rocks, chiefly greywackes. The Cumberland Bay formation, a unit of volcanic greywackes composed of interbanded shale and sandstone, differs geochemically from the Sande-bugten formation, a unit of quartzose greywackes which contain a higher ratio of sandstone and shale (Clayton, 1982). Igneous rocks are restricted to the SW corner of the island (Bell, Mair & Storey, 1977).

The island has an alpine topography, much of it lying over 1000 m, and about 60% of its surface area is covered by permanent ice which reaches the sea in many large glaciers at the heads of fjords (Clapperton, 1971; Figure 3.2A). The island lies 200 km south of the Antarctic Convergence and the surrounding cold water mass (Figure 3.2B) results in a cool oceanic climate typical of subantarctic islands. Winter and summer seasons are very clearly defined on South Georgia (Figure 3.3), although the annual range of mean monthly temperatures is only about 7 °C (Figure 3.4). However, sub-zero mean monthly air temperatures occur for four to five months of the year and the absolute range is −19 to 24 °C at sea level. There are prevailing westerly winds, annual precipitation is high (1200–2000 mm), summer snowfalls are frequent and there is deep snow cover at sea level for half the year (Figure 3.3B).

The vascular flora of South Georgia is poorly developed because of the island's isolation from any major landmass and the extension northwards of the Antarctic Convergence to produce polar–alpine conditions at a relatively low latitude (Lewis Smith, 1984). During the Pleistocene glaciation, South Georgia was completely covered by ice which began to recede 10–12 000 years ago (Sugden & Clapperton, 1977). Pollen analyses of peat have confirmed that plant communities developed since that time, with wind and birds playing a major role in seed dispersal (Lewis Smith, 1984). The native flora comprises 26 species (10 monocotyledons, 9 dicotyledons and 7 pteridophytes). In contrast, the cryptogamic flora is much more diverse and comprises approximately 175 species of moss, 85 liverworts and 160 lichens (Lewis Smith, 1984). Relatively few species achieve dominance and only eight principal community types have developed (Lewis Smith & Walton, 1975), some of which are illustrated in Figure 3.5 and which will be described further in Chapter 5. The tussock grassland community is dominated by *Poa*

Figure 3.1. Simplified geological map of South Georgia (after Bell, Mair & Storey, 1977), also showing areas occupied by reindeer herds.

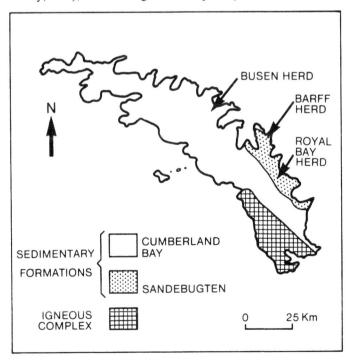

Figure 3.2. South Georgia has an alpine topography dominated by Mt Paget (2965 m); 60% of its surface area is covered by permanent ice and snow, mainly in the form of heavily crevassed glaciers entering the sea at heads of fjords, the largest of which is the Nordenskjöld Glacier lying to the west of the Barff Peninsula (A). The island lies south of the Antarctic Convergence and is surrounded by cold water (B) which has a considerable effect upon its climate.

Figure 3.3. Summer and winter seasons are well defined; the permanent snow level in summer lies between 100 and 200 m (A) and deep snow cover lies at sea-level during winter (B). Stromness whaling station lies in the background.

flabellata, recently redescribed in a new monospecific genus *Paro-diochloa* (Groves, 1981). I have retained the old name of the genus throughout this monograph, pending wider acceptance and knowledge of the revision. *P. flabellata* is the climax vegetation and forms the most widespread and productive community on South Georgia (Lewis Smith, 1984).

The terrestrial fauna is also poorly developed. Most invertebrates have a negligible effect upon the vegetation in terms of herbivory (Lewis Smith & Walton, 1975). The dominant herbivores which occur on South Georgia are three species of arthropod which eat the base of grass shoots and act as decomposers (Gressitt, 1970; Vogel, 1985). There are no obligate terrestrial carnivores and apart from an endemic species of pipit and duck which depend on terrestrial food resources (Burger, 1985), all

Figure 3.4. Climate of South Georgia compared with an alpine tundra area in south Norway (data from Lewis Smith & Walton, 1975; Østbye *et al.*, 1975).

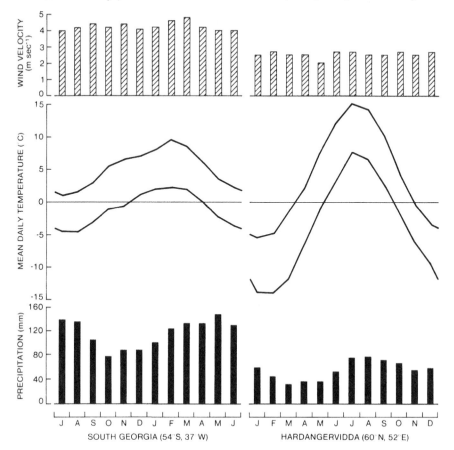

vertebrates found on the island depend upon the considerable resources of the marine ecosystem for their food. Thus the large colonies of seabirds, including penguins and albatrosses and fur and elephant seals only come ashore to nest and breed (Croxall, 1984; Laws, 1984). However, they do have a considerable effect upon the vegetation, enriching it around their rookeries and colonies, and even damaging it by trampling in densely populated areas (Walton & Lewis Smith, 1980; Bonner, 1985). The skua and Dominican gull are the only predatory birds and feed on small petrels and around penguin rookeries; they are also scavengers together with sheathbills and giant petrels.

South Georgia and other subantarctic islands are tundra habitats, but differ greatly from the many northern tundra regions of which reindeer and caribou are native (French & Smith, 1985). The climate of southern tundra regions is more stable, but the more favourable temperatures are well compensated for by high winds (Figure 3.4). The floras of southern tundras are poor in species and lack any truly woody shrubs (*Acaena* spp. are suffruticose herbs and have woody stems and rhizomes only: Lewis Smith, 1984), as Captain Cook rightly observed when bemoaning his lack of a toothpick. However, southern tundra has a much more luxuriant vegetation, and has favoured species with slow but consistent rates of reproduction compared with northern species which are adapted to short growing seasons (Callaghan, Lewis Smith & Walton, 1976; French & Smith, 1985). Most southern floras are only adapted to grazing by arthropods, as terrestrial mammals are not native to these islands. Finally, the lack of predators in southern regions was one of the most obvious differences that reindeer faced when introduced to South Georgia.

3.2   History and introduction of animals

The discovery of South Georgia by Captain Cook in 1775, and the descriptions of the abundant biological resources by the naturalists aboard the *Resolution*, soon brought voyagers to the island (Matthews, 1931; Headland, 1984). There was extensive exploitation of the fur seal during the 19th century, which nearly resulted in its extermination. However, protection afforded to the species early this century has resulted in a dramatic recovery of fur seals around the island (Payne, 1977; Bonner, 1982). The Antarctic whaling industry began at South Georgia with the opening of the Grytviken station in the 1904/5 season. Several other stations were established between then and 1920, and the last one was closed in 1964/5 following the decline of the whaling stocks

Figure 3.5. Vegetation of South Georgia in areas not grazed by reindeer, showing (A) coastal tussock grass, *Poa flabellata*, and a stand of greater burnet, *Acaena magellanica*; (B) oligotrophic mire dominated by the rush, *Rostkovia magellanica*; (C) the macrolichen, *Pseudocyphellaria endochrysea*, in dry meadow; and (D) fellfield dominated by the grasses, *Festuca contracta* and *Phleum alpinum*.

(Headland, 1984). An elephant sealing industry flourished during the period of whaling and provided an unusual example of rational management under government control (Laws, 1953).

South Georgia, now part of the Crown Colony of South Georgia and the South Sandwich Islands, was originally a Dependency of the Falkland Islands and since 1909 has had a resident Magistrate and Administrative Officer. Following the termination of the whaling and sealing industries for economic reasons, the administrative settlement on King Edward Point was taken over by the British Antarctic Survey as a research station in 1969. It was from there that I carried out my fieldwork between 1973 and 1976. The name of the station was changed to Grytviken after the nearby disused whaling station in 1977.

As the subantarctic islands lacked indigenous terrestrial mammals, attempts were made to introduce and maintain herbivorous animals and birds upon them (Holdgate & Wace, 1961; Bonner, 1984a; Leader-Williams, 1985). The first introduction to South Georgia was of rabbits in 1872 and at various times there have been introductions of horses, cattle, sheep, goats, pigs and upland geese (Allen, 1920; Matthews, 1931; Headland, 1984). None of these has become established on the island as a feral population. Cows were kept for milk, goats for milk and meat, and sheep and bullocks for summer fattening. Domestic stock was normally housed in winter, and pigs usually housed year round; sheep occasionally survived winters unhoused. Rodents have been introduced accidentally. Brown rats are established on many lowland areas of the island. Many separate introductions were probably made by sealers living in camps around South Georgia (Olstad, 1930; Pye & Bonner, 1980). There is also an isolated population of house mice on the NW side of the island in an area where rats are not present (Bonner & Leader-Williams, 1977; Berry, Bonner & Peters, 1979). Carnivores are not established on the island. Dogs and cats have been introduced as pets, and an unsuccessful fur industry was attempted based on silver foxes. Additionally there have been about 64 species of vascular plants introduced to the island, of which only a few still persist and occur largely in the vicinity of the whaling stations (Walton, 1975; Headland, 1984).

On three occasions between 1911 and 1925, small numbers of reindeer were introduced by the Norwegian whalers to two geographically isolated areas of the island in the vicinities of whaling stations (Figures 3.6A, 3.7). These introductions presumably aimed to provide the island with a sporting amenity (Figure 3.6B) and alternative source of fresh meat that required no husbandry or management. The fates of the three introduc-

Figure 3.6. Reindeer were introduced to South Georgia by Norwegian whalers who occupied shore-based stations (A) on the island from 1904 to 1965: the reindeer were introduced primarily as a sporting amenity (B). (Lower photograph reproduced from *Norsk Hvalfangst-Tidende* **44**(2), 88, 1955.)

tions differed greatly and Leader-Williams (1978) attempted to clarify inconsistencies between the various reports (Allen, 1920; Olstad, 1930; Bonner, 1958; Lindsay, 1973; Kightley & Lewis Smith, 1976) and summarise archival information (Falkland Islands Dependency of South Georgia File No. 650) to produce a coherent picture of the history of each introduction.

The first introduction was made into New Fortuna Bay (renamed Ocean Harbour in 1954) on the Barff Peninsula by managers of two of the whaling stations, the brothers C.A. and L.E. Larsen. They had bought 11 reindeer at Valders, a valley in Numedal, in the central part of southern Norway for shipment to South Georgia. One apparently died on the voyage, for it seems that only 10 reindeer (three males and seven females) were liberated in November 1911. The second introduction was made during the same whaling season of 1911/12 by another whaling company into Leith Harbour, Stromness Bay. However, the five reindeer introduced increased in number to about 20, but then all perished in a snow-slide (Olstad, 1930). The third introduction was made into Husvik Harbour, also in Stromness Bay, in 1925 and comprised seven reindeer

Figure 3.7. Map of South Georgia showing numbers of reindeer introduced with dates and places of introduction. Stippled areas indicate areas of permanent ice and snow (after Leader-Williams, 1978).

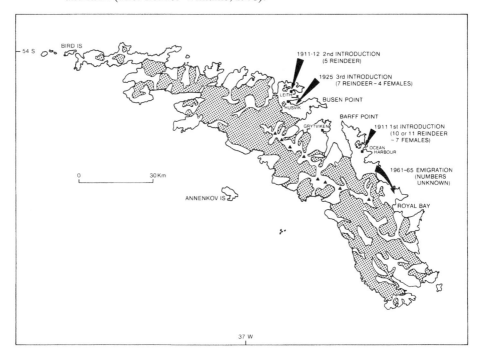

(three males and four females). All efforts to trace the place of purchase in Norway have been unsuccessful. Since their introductions, there has never been any contact or interchange of reindeer between the two areas, because the Barff Peninsula and the Stromness Bay or Busen (hereafter referred to as Busen) area are well separated by glaciers which act as barriers to movement (Figure 3.7).

Some of the female reindeer were pregnant when transferred to South Georgia, and gave birth in May 1912, but many calves perished at the onset of the austral winter (Olstad, 1930). It seems that female reindeer took two years to reverse their breeding season, and gave birth to the first calf conceived on South Georgia in November 1913. This reversal in breeding season and in other events of the annual cycle (Figure 3.8) by

Table 3.1. *Historical summary of reindeer numbers on South Georgia (from Leader-Williams, 1978)*

Barff herd		Busen herd	
Nov. 1911	Import of 10 deer – 3♂, 7♀	1911–12	Import of 5 deer – 2♂, 3♀
May 1912	6 calves born (Olstad, 1930)	May 1912	3 calves born (SG file 650)
Nov. 1913	Birth of first calf crop conceived on South Georgia		
Mar. 1916	45 deer total (SG file 650)		
		1917	17 deer total (SG file 650)
1920	*c.* 120 deer (SG file 650)		
		?	All killed in snowslide (Olstad, 1930)
1921–2	Nearly 300 deer (Wilkins, 1925)		
		1925	New import 7 deer – 4♂, 3♀ (Olstad, 1930)
1928	400–500 deer (Olstad, 1930) +150–200 killed by hunters		
1953	2000 deer (SG file 650)	1953	40 deer (SG file 650)
1955–7	4000 deer (Bonner, 1958) (range 3000–5000)		
		1955–7	100–200 deer (Bonner, 1958)
1961–5	Spread to Royal Bay (Lindsay, 1973)		
1972	Initiation of British Antarctic Survey Programme		
1972	1300 deer Barff +800 deer Royal Bay (Table 3.2)		
		1973	800 deer (Table 3.2)

Figure 3.8. (A) Reindeer calves on South Georgia are born in the austral spring, during November, which is a reversal of six months compared with their progenitors in Norway, and other events in the annual cycle such as the moulting of winter coats (B) are similarly affected.

six months is discussed further in Chapter 4. Information on numbers of reindeer present since their introduction is anecdotal and fragmentary, and it was not until 1972 that accurate aerial counts were undertaken (Table 3.1). Surprisingly, however, it appears that a realistic reconstruction of events can be made when all the available estimates and counts are pieced together (Figure 3.9).

It seems the 10 reindeer introduced to the Barff Peninsula increased rapidly and by 1958 numbered about 3000 animals (Leader-Williams, 1978). The reindeer were protected by legislation from 1912, and the first permit was issued to the whalers in 1916 to shoot males only (Figure 3.6B). From then until 1930, about 150 to 200 deer had been harvested under license, and from the 1930s to 1950s up to 100 animals were taken annually from this herd (Olstad, 1930; Leader-Williams, 1978). During the 1950s whales were becoming scarce and it was no longer economic to deploy a whale catcher for a day's reindeer hunting. From 1955 until 1972 the numbers of deer taken therefore dwindled to about 20 per year (Bonner, 1958; Leader-Williams, 1978). Since the 1950s there have been marked changes in the Barff herd. First, it seemed there was a decline in numbers which cannot be attributed to overhunting. Secondly, there was a considerable increase in the extent of overgrazing, especially of the normally abundant coastal tussock grass (Kightley & Lewis Smith, 1976). Thirdly, part of the herd spread across or in front of the Cook Glacier (Figure 3.10), probably between 1961 and 1965, to form a new herd at Royal Bay (Lindsay, 1973). By 1972 the Barff herd numbered 1300 and the Royal Bay herd 600 reindeer (Table 3.2). The other herd in the Busen area was introduced near whaling stations and its increase was restricted probably by poaching during the 1940s. In the 1950s protection became more effective, and the Busen herd began to increase (Bonner, 1958), numbering 800 reindeer by 1972. Thus at the start of my study there were two genetically distinct stocks of reindeer which formed three herds.

3.3 The study

The present study was initiated in response to the concern expressed by British Antarctic Survey botanists at the apparent increase in the extent of overgrazing observed since Bonner's (1958) study and at the recent occupation of a new area of the island by reindeer (Lindsay, 1973; Kightley & Lewis Smith, 1976). A preliminary study was undertaken by M.R. Payne in the austral winters of 1972 and 1973, and I undertook fieldwork between November 1973 and March 1976. Because

Figure 3.9. Estimates of reindeer numbers in different herds on South Georgia (data from Tables 3.1, 3.2). Solid line = total number of reindeer; broken line = numbers of adults and yearlings; arrow = formation of Royal Bay herd; star = combined total from Barff and Royal Bay herds (from Leader-Williams & Payne, 1980).

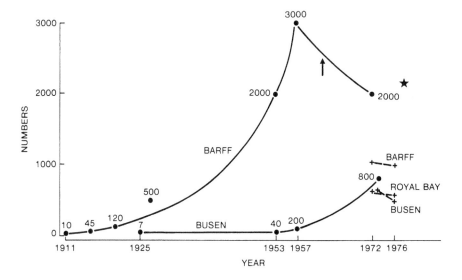

Figure 3.10. The Cook Glacier has a narrow snout ending on a foreshore. It lies to the south of the Barff Peninsula and gave reindeer access to the Royal Bay area in 1961–5.

of the effect of reindeer on the vegetation, and the possible need for management, it was felt there was sufficient justification to cull reindeer for the programme, and the necessary permits were issued under the Falkland Islands Dependencies Conservation Ordinance by the Magistrate of South Georgia. The results described in the following chapters are based mainly upon the following samples: 300 reindeer (113 males and 187 females) from the Barff herd shot in all months between May 1974 and November 1975 except August and October 1974; 100 reindeer (42 males and 58 females) from the Royal Bay herd collected in only two eight-week periods, from mid-December 1973 to mid-February 1974 and during the same months two years later; 100 reindeer (39 males, 60 females and one intersexual) from the Busen herd collected during six different months between August 1974 and February 1976; and 146 carcasses resulting from natural deaths collected from all three herds.

Methods used to process the samples will be described only in sufficient detail in each chapter for the reader to understand the results. For those interested, the bibliography will provide original sources for further consultation. One technique central to the study was the assignment of age to all specimens. This was achieved by a combination of tooth eruption patterns and annulations in decalcified sections of incisor teeth, and an example of the latter is shown in Figure 3.11 (Leader-Williams, 1979a). For future reference the age structure of specimens collected is shown in Figure 3.12. Before results from the study can be considered in detail, it is necessary to define the stage that each herd had reached in their irruptive oscillation (Figure 2.1).

Table 3.2. *Herd sizes estimated from total aerial counts. Numbers of calves were not recorded (NR) in all counts (from Leader-Williams, 1980b)*

Date	Barff herd Adults and yearlings	Calves	Royal Bay herd Adults and yearlings	Calves	Busen herd Adults and yearlings	Calves
Dec. 1972	1050	180	630–640	135–145		
Nov. 1973					639	146
Jan. 1976			530–570	NR	450–480	NR
Dec. 1976	1000	320				

Figure 3.11. Longitudinal section through cementum of an incisor both showing dark rest lines (RL) each of which is laid down annually. The age of this reindeer was determined as 7½ years. Scale bar represents 0.01mm (from Leader-Williams, 1979a).

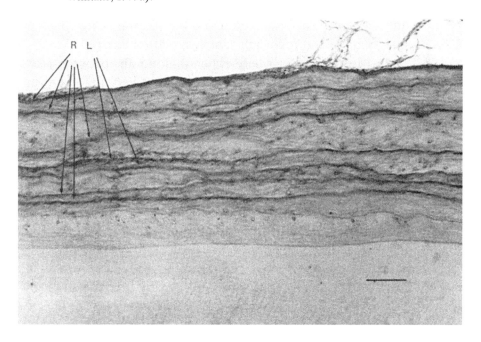

Figure 3.12. Age distribution of shot reindeer in different herds shown as a percentage of sample size. The star indicates that one intersex reindeer was not included in the total for the Busen herd.

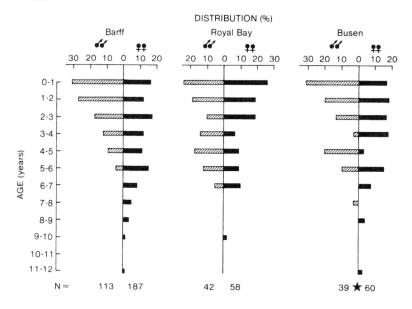

3.4   Status of reindeer during the study

Assessment of the status of each herd relied upon having information on numbers of reindeer (Table 3.2, Figure 3.9), the area of range occupied by each herd (Figure 3.13) and their densities (Table 3.3).

Aerial counts were made in early summer between 1972 and 1976 by oblique aerial photography of reindeer groups encountered during systematic searches with two helicopters; these were calibrated by total counts on foot and found by M.R. Payne and myself to be accurate and to within 10% of each other for adults and yearlings (Leader-Williams, 1980b; Leader-Williams & Payne, 1980). The localities used by reindeer were assessed on foot and the areas occupied were traced from monochrome aerial photographs taken in November 1973 onto maps. Total areas available to reindeer and the area of range covered with vegetation and occupied by them were measured with a planimeter (Leader-Williams, 1980d).

Figure 3.13. Areas of South Georgia occupied by reindeer herds. Stippling = area of permanent ice and snow; hatching = areas of vegetation occupied by reindeer; solid = areas within glacier boundary of Busen herd not occupied by reindeer in 1976.

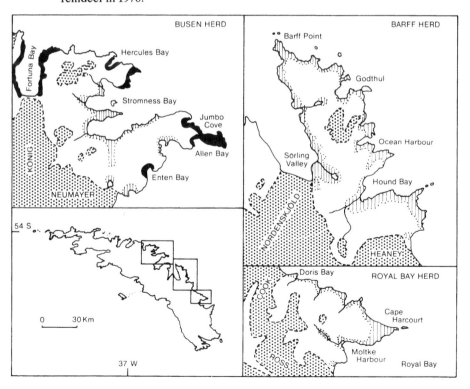

Numbers of reindeer in the Barff herd remained steady during the study at around 1000 adults and yearlings, even though 365 reindeer were shot for the preliminary, and for my own, project between December 1972 to 1976. The Barff herd was the first to be introduced to South Georgia, and was at an advanced stage in its irruptive oscillation. An inspection of Figure 3.9 suggests that the herd might either have been in the decline stage, or have recently entered the post-decline stage, but the data available between 1957 and 1976 do not allow resolution of this point. The Barff herd occupies and grazes all accessible areas within its range (Figure 3.13) apart from impossibly steep sea and inland cliffs. The herd still occurs at very high densities compared with established populations in the Northern Hemisphere (Figure 3.14).

Numbers in the Royal Bay herd also appeared to remain steady during the study (remembering that there was a probable error of 10% in counting methods), even though 100 reindeer were shot for the study. This herd also grazes all its accessible range, and densities in 1976 were similar to peak densities calculated for the Barff herd (Table 3.3). In spite of the recent spread of reindeer to this new area, overgrazing of lichens, *Acaena* spp. and tussock grassland was already evident (Lindsay, 1973; Kightley & Lewis Smith, 1976). This information suggests that quite a large number of reindeer moved into the Royal Bay area, and that the herd was approaching or had already reached maximum numbers at the start of the study. Visits to the Cook and Heaney glaciers

Table 3.3. *Areas of range available and used, densities of reindeer and summary of differences in status of each herd on South Georgia during 1972–6 (from Leader-Williams, 1980d)*

Parameter	Herd		
	Barff	Royal Bay	Busen
Date introduced	1911	1961–5	1925
Total area (km^2)	131	58	124
Peak densities (animals km^{-2})	23	13	6
Vegetated area occupied by reindeer (km^2)	30	9	11
Peak densities in occupied and grazed areas (animals km^{-2})	100	85	71
1976 densities in occupied and grazed areas (adults and yearlings km^{-2})	33	61	42
Probable stage in irruptive oscillation	3/4	1/2	3
Genotype	1	1	2

made by myself and other field parties suggested that interchange between Barff and Royal Bay herds no longer occurred.

In contrast to the other two herds, numbers in the Busen herd declined sharply during the study (Table 3.2) even though only 100 animals were shot. Inspection of Figure 3.9 suggests that the Busen herd had entered the decline stage of its irruption. The total area available to the Busen reindeer between the Fortuna and Neumayer glaciers is similar to the total area of the Barff Peninsula, but in 1976 the Busen reindeer had not yet negotiated four mountain passes leading to potentially available range (Figure 3.13), even though other areas of their range were overgrazed (Kightley & Lewis Smith, 1976). The reasons for this difference between the two stocks is unclear, for the mountain passes on the Busen Peninsula appeared no more steep or difficult than those crossed regularly by Barff and Royal Bay herds. A possible reason may be that the two stocks originated from different areas of Norway, the Barff introduction being of mountain stock and the Busen introduction being perhaps of more sedentary forest type. In spite of behavioural differences, however, when densities are calculated in terms of areas grazed and occupied, peak and 1976 densities between the Barff and Busen herds were very similar (Table 3.3).

The chapters in Part II will discuss data collected from the Barff herd,

Figure 3.14. Examples of densities attained by wild and introduced herds of reindeer and caribou. (Data are shown only for the Barff herd on South Georgia. Those for other herds are from sources listed in Figures 1.8 and 2.6, and from Bergerud, 1980a; Strandgaard *et al.*, 1983; Skogland, 1985.)

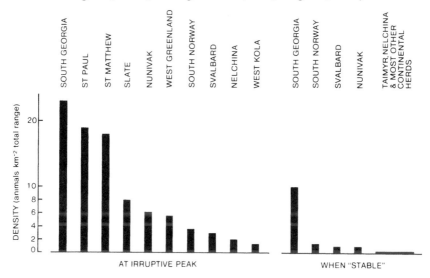

except where otherwise stated. More detailed comparisons between the three herds, and differences between the irruptions on South Georgia and on other islands, will be discussed in Part III.

3.5   Summary

1. South Georgia is a subantarctic island with a species-poor terrestrial flora and fauna.
2. Norwegian reindeer were introduced to the island on three occasions between 1911 and 1925. There are now three herds comprised of two genetic stocks, and each at a different stage of their irruption.
3. The study is based mainly upon reindeer shot between 1973 and 1976.

Part II

Biology of reindeer

This section describes the comparative biology and ecology of reindeer. *Rangifer* are amongst the more sexually dimorphic species of temperate cervid and generally live in large groups in open habitat. The reindeer of South Georgia have made adjustments in adapting to their unusual environment, which is characterised by a limited and novel flora and by a lack of predators. An early adaptation was reversal of their breeding season, so that they give birth in the austral spring. On South Georgia, as in the Northern Hemisphere, male and female reindeer show differences in many aspects of their life histories and this provides an opportunity for a comparative study of sexual dimorphism in *Rangifer*. Male and female reindeer show differences in their reproductive physiology, including the age of attainment of puberty and the control of their antler cycles (Chapter 4). The forage available for reindeer on the island differs from that in the Northern Hemisphere, but factors underlying forage selection are similar. Surprisingly, there are few differences in the feeding ecology and nutritional status of males and females (Chapter 5). However, patterns of growth and fat deposition show marked differences in males and females. Periods of weight loss are related to the main periods of reproductive expenditure in each sex (Chapter 6). Disease and parasites are important contributory causes of mortality on South Georgia, but their prevalence does not differ in males and females. Loss of body fat and sexual dimorphism in body growth are important determinants of the patterns of mortality in males and females (Chapter 7). *Rangifer* and other northern cervids are amongst the most widely studied of the large herbivorous mammals. Throughout the section, data from reindeer on South Georgia are compared with results derived from studies in the Northern Hemisphere.

4

Reproduction

The rutting season lasts but a fortnight, during which period the male is savage
and dangerous. Immediately afterwards he casts his horns.
Carl Linnaeus (1732)

4.1   Introduction

The survival of any species living in a seasonal environment
depends on the ability to produce its young at favourable times of year.
Environmental factors have an ultimate influence on the timing of
breeding, for they define the optimum period of birth (Baker, 1938).
However, such factors are discontinuous and variable in their onset each
year, and animals have adopted cues that are related to highly predic-
table changes in the solar system to determine the timing of their
breeding season. Furthermore, the duration of the gestation period is
fixed in most mammals, excepting those which show delayed implan-
tation, and control of the time of birth actually requires control of the
time of conception. In the case of temperate and polar ungulates,
reproductive activity has become entrained on annual changes in day
length (photoperiod) which constitute the proximate cue to the timing of
the mating season (Lincoln & Short, 1980). The optimal time for birth is
in spring, when weather conditions ameliorate and several months of
plant growth lie ahead.

Reindeer and caribou mate in the autumn, when females come into
oestrus, ovulate and are sexually receptive for periods of up to 2 days.
They are classed as a seasonally polyoestrous species because successive
oestrous cycles will occur only within a limited breeding season if mating
does not result in pregnancy. After a successful mating the fertilised egg
is implanted without delay and single young are born in spring after the
gestation period lasting between 208 and 240 days.

In this chapter, I first examine the timing of the breeding season on

South Georgia in relation to other populations of reindeer and caribou (4.2). Then I examine pregnancy rates and ovarian changes in females (4.3) before discussing reproductive changes in males of different ages (4.4) and the timing of the antler cycles in males and females (4.5). Finally, I compare the breeding biology of reindeer with other temperate species of cervid (4.6). I use data in this chapter to examine the adaptations of *Rangifer* to life at high latitudes, and to determine whether there is a dimorphism in the reproductive performance of males and females.

4.2   The breeding season

Marshall (1937) investigated various factors influencing the onset of the breeding season and cited the example of red deer that had reversed their oestrous cycles by 6 months after being transported from Europe to New Zealand, and then showed a further reversal after being transported back again. This study determined the crucial role that changes in photoperiod play in the onset of breeding seasons. A similar reversal was shown by reindeer transported to South Georgia (Olstad, 1930).

4.21   The rut

Rutting behaviour of *Rangifer* males is characterised by cessation of feeding, tending of females, hunching and urinating on the hind feet, panting, bush-thrashing, fighting with rival males and, of course, mating (Espmark, 1964a; Lent, 1965a; Henshaw, 1970; Bergerud, 1973, 1974a; Thomson, 1977). Apart from observing changes in group composition and size, I did not make a systematic study of rutting behaviour. However, qualitative observations provide an adequate description of the timing of the rut on South Georgia. Some, but not all, reindeer males remain segregated from females for most of the year, but mixed herds are encountered at all seasons (Figure 4.1). As the rutting season approaches herds are almost entirely mixed and become reduced in size. At the rut, median group size is 15 to 20 reindeer, reduced gradually throughout summer from large group sizes of over 100 in spring.

92 groups of reindeer were classified systematically at different seasons of the year according to their composition (of mixed sex, males only, and female/calf groups) and size. In the case of large groups >50 only approximate group sizes were assigned. For analysis, groups of 1–10, 11–50 and >50 were combined; group composition ($\chi^2 = 21.37$, $df = 6$, $P < 0.005$) and group size ($\chi^2 = 25.64$, $df = 6$, $P < 0.001$) change at

different seasons. Animals in small parties are encountered more frequently than those in larger ones and the size of the group in which the average animal finds itself is more meaningful than mean group size (Clutton-Brock *et al.*, 1982). Therefore I estimated median group size from a cumulative frequency table constructed from the numbers of individual reindeer found in groups of different sizes.

Rutting behaviour is first seen during the last week of February when, as a substitute for bush-thrashing, reindeer on South Georgia begin to 'tussock-thrash'. Antler racks are rubbed vigorously through large tussocks of *Poa flabellata* for periods lasting up to 30 minutes, and a tussock so mistreated has its leaves strewn far and wide over a 5 m radius! Sparring between males is seen in the first week of March, and protracted fights, tending of females, panting and attempted mountings of females are seen in the last two weeks of March (Figure 4.2). These rutting behaviours continue during the first two weeks of April, after which the rut loses its intensity and fights become sporadic.

I observed no successful matings during daylight hours, but the period of the rut can be defined approximately as the last two weeks of March

Figure 4.1. Changes in group size and composition of reindeer herds on South Georgia. Sample sizes are shown as appropriate.

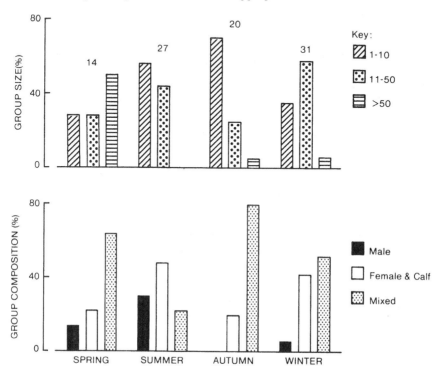

and the first two weeks of April. Initially the rut involves only prime adult males (see also Figure 6.1A), and it is unlikely that an individual male is involved for the whole period of the rut. 'Rutted-out' males, exhausted and playing no further part in the rut (see Figure 6.1C) are observed as early as the last week of March, and by the end of the rut only young adults and yearling males show signs of rutting behaviour, and spar with each other. A study of individually recognisable, free-living *Rangifer*, undertaken in Newfoundland, showed that a dominance hierarchy existed amongst males and that prime males were involved in rutting activities for only about 10 days (Bergerud, 1973). My observations from South Georgia support this suggestion of a short and intense rut.

4.22   *Oestrus and calving*

The oestrous cycle of reindeer and caribou varies in length and can either be short (10 to 12 days) or long (24 days) (McEwan & Whitehead, 1972; Dott & Utsi, 1973; Bergerud, 1975). The length of oestrus, when the female is sexually receptive and ovulation occurs, varies from 1 to 2 days (Gorbunov, 1939; Bergerud, 1975). A 'silent' oestrus is said to occur at the start of the breeding season, when an ovulation takes place but there is no overt display of oestrous activity (Bergerud, 1975; McEwan & Whitehead, 1980). On South Georgia, the earliest observation of an ovulation in a shot reindeer was made on 22 March. Furthermore, this and all other subsequent observations of first ovulations ($N = 4$) were accompanied by evidence of an overt oestrus and a recent mating. Therefore the occurrence of a silent oestrus was not confirmed on South Georgia. Cumulative evidence of recent ovulations in shot reindeer shows that the mean date of oestrus is 7 April with an approximation to the true standard deviation of ± 2 days (Figure 4.3). If the first oestrus is overt, this suggests that 95% of conceptions occur between 3 and 11 April. The restricted period in which most matings apparently occur coincides with the qualitative observations of rutting behaviour in males.

As expected from the short period of the rut, the season of births in *Rangifer* is similarly restricted, both in captivity (Zuckerman, 1953) and in the wild (Table 4.1). The first calf born in each year on South Georgia was seen on 30 October 1974 and 31 October 1975. The calculated mean date of birth is 20 November ± 6 days, suggesting that 95% of calves are born between 8 November and 2 December. The interval between the mean dates of oestrus and of birth, each calculated from independent

Figure 4.3. Mean dates of oestrus and calving calculated from number of reindeer with luteal tissue (*C*) and that had calved (*L*), divided by total sampled for each collection period (*B*, *B'*). Sample sizes are shown above each column, and mean dates are shown with approximate standard deviations.

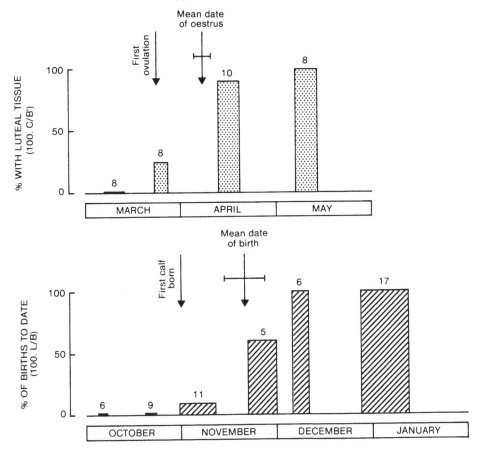

Figure 4.2. (left) Rutting behaviour lasts from mid-March until mid-April when (A) males fight rivals, (B) herds become mixed and males attempt to maintain for themselves a group of females, and (C) oestrous females stand for successful males, which commonly hunch and urinate on their feet.

Table 4.1. *Main calving seasons of reindeer on natural range, introduced reindeer and caribou in different localities, showing periods when approximately 90–95% of calves are born*

Main calving season[a]	Sub-species or type	Location	Latitude/longitude	Source
? April – ? May	Herded reindeer	China	52°N 120°E	Ma Yi-ching (1983)
1 May – 15 May	Herded reindeer	USSR – Europe	65°N 55'E	Zhigunov (1969)
6 May – 31 May	Wild reindeer	Norway – South	62°N 9°E	Holthe (1975); Thomson (1977)
8 May – 21 May	Herded reindeer	Sweden – Lapland	67°N 21°E	Espmark (1971a)
10 May – 29 May	Herded reindeer	Finland	69°N 27°E	Eloranta & Nieminen (1986)
22 May – ? June	Forest reindeer	Finland/USSR – Karelia	63°N 30°E	Helle (1981)
2 June – 9 June	Spitzbergen reindeer	Svalbard	78°N 20°E	Tyler (1987a)
16 June – 3 July	Tundra reindeer	USSR – Taimyr	75°N 100°E	Michurin (1967)
13 April – 9 May	Introduced reindeer	Canada – Mackenzie Delta	69°N 134°W	Nowasad (1975)
? April – ? May	Introduced reindeer	Alaska	66°N 164°W	Naylor et al. (1980)
? May	Introduced reindeer	Scotland	57°N 4°W	Lindgren & Utsi (1980)
8 May – 2 June[b]	Introduced reindeer	South Georgia	54°S 37°W	This study
21 May – 31 May	Introduced reindeer	Iceland	65°N 16°W	Thórisson (1983)
24 May – 26 May	Alaskan caribou	Alaska – Central	62°N 150°W	Skoog (1968)
? May – ? June	Woodland caribou	Canada – Alberta	57°N 112°W	Fuller & Keith (1981)
24 May – 2 June	Woodland caribou	Canada – Newfoundland	48°N 56°W	Bergerud (1975)
4 June – 15 June	Barren-ground caribou	Canada – NWT	63°N 94°W	Parker (1972); Miller & Broughton (1974)
5 June – 7 June	Woodland caribou	Canada – Labrador	55°N 65°W	Muller-Schwarze & Muller-Schwarze (1983)
3 June – 11 June	Barren-ground caribou	Greenland – West	67°N 53°W	Thing (1981)
8 June – 13 June	Alaskan caribou	Alaska/Canada	72°N 152°W	LeResche (1975)
? June	Peary caribou	Canada – Arctic Islands	75°N 100°W	Thomas (1982)

[a] Variation in data does not allow precise definition of calving season in many instances.

[b] Reversed by 6 months.

Table 4.2. *Gestation periods of reindeer and caribou in different localities*

Average gestation length (days)	Species or type	Locality	Source
208	Introduced reindeer	Canada (captive)	McEwan & Whitehead (1972)
214	Introduced reindeer	Scotland	Dott & Utsi (1973)
216	Barren-ground caribou	Canada (captive)	McEwan & Whitehead (1972)
223–7	Herded reindeer	USSR	Schmitt (1936); Gorbunov (1939); Preobrazhenskii (1969)
[229]	Introduced reindeer	South Georgia	This study
229	Woodland caribou	Canada – Newfoundland	Bergerud (1975)
220–40	Herded reindeer	Alaska	Hadwen & Palmer (1922); Palmer (1934)

data sets, is 229 days. Whilst this subtraction cannot be considered as a true measure of gestation length, it is nonetheless similar to data for *Rangifer* elsewhere (Table 4.2).

> Calculations of the mean dates of oestrus and of calving were made using a modification of the Spearmann-Karber method (Finney, 1964) for the analysis of quantal response data in biological assay, as described by Laws, Parker & Johnstone (1975) for estimating the mean age of puberty. Data for the mean date of oestrus were based on the number of females with an active corpus luteum of oestrus or of pregnancy, indicating a recent ovulation (C), divided by the sampled total of sexually mature females (B'). These data resulted from sectioning of ovaries, described in Section 4.3. Data for mean date of birth were derived from the number of post-partum females (L), divided by the sampled total of post-partum and pregnant females (B), shot during periods shown in Figure 4.3.

The timing of the calving season differs in *Rangifer* populations (Table 4.1). By analogy with mountain sheep, calving might be expected to begin later and to be of shorter duration at more northerly latitudes (Bunnell, 1982). Unfortunately, data for *Rangifer* are of variable quality and insufficient for a detailed analysis of factors affecting timing of the calving season. In contrast to mountain sheep, *Rangifer* at all latitudes have a restricted calving season. However, the timing of the calving season of reindeer on natural range is delayed at northerly latitudes (Figure 4.4), as in mountain sheep (Bunnell, 1982). A more extensive data set for caribou would also probably show this trend (Table 4.1). Bunnell (1982) related timing in mountain sheep to the effect of phenological differences in vegetation growth on maternal energy requirements, and a similar mechanism is likely in *Rangifer*.

Caughley (1971) examined timing of the reversal of calving in ungulates introduced to New Zealand and found anomalies between southern and northern hemispheres, probably because northern data were then only available for London Zoo (Zuckerman, 1953). Furthermore, no clear latitudinal differences are evident in the timing of the calving season in zoo populations of red deer (Fletcher, 1974). Introduced reindeer also appear to show anomalous timing of their calving seasons, which do not relate to their present latitudes (Figure 4.4). The most extreme of these is the Mackenzie Delta herd that calve 4 to 6 weeks before native caribou occupying adjacent range. However, examination of the timing of the calving season in areas of their origin shows that the North American introduction has an early calving season similar to reindeer in extreme eastern areas of USSR (Zhigunov, 1969). The timing of the calving season in Icelandic and South Georgian reindeer is also similar to present-day Norwegian reindeer, apart from the six month reversal in the Southern Hemisphere (Table 4.1). Thus introduced reindeer have retained the calving season of their area of origin. This suggests that timing of the breeding season may have a genetic basis and that it requires considerable time to adapt to differences in phenology in the area of introduction.

Figure 4.4. Comparison of timing in calving peaks of reindeer at different latitudes, for both natural and introduced populations. Calving date is presented as mid-date for main calving period (data from Table 4.1).

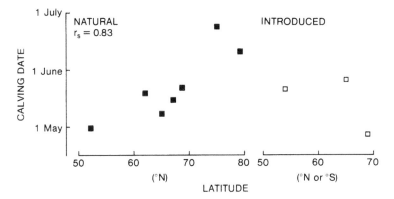

4.3   Pregnancy rates and reproductive changes in the female

The ovary contains many thousands of primordial follicles at birth, but only a few hundred of these develop into primary follicles (Harrison, 1962). Around the time of puberty, primary follicles develop in sequence and enlarge rapidly to secrete the basic sex hormone, oestrogen. Ovulation takes place when a follicle develops to maturity (Figure 4.5A) and ruptures to release its ovum. Cells then grow from the outside of the old follicle to form a corpum luteum (referred to hereafter as CL), a solid structure which secretes another hormone, progesterone. If pregnancy does not ensue, the CL will survive for only a few days as a CL of ovulation, before degenerating to allow another follicle to develop to maturity and ensure a subsequent ovulation. When an ovum is fertilised and reaches the uterus, the CL of many species, including cervids, persist to secrete progesterone which maintains pregnancy (Figure 4.5B). Twinning is unusual in reindeer and caribou (McEwan, 1971; Nowasad, 1973; Godkin, 1986), but several ovulations can occur and result in the formation of secondary CL (Figure 4.5B,C). After

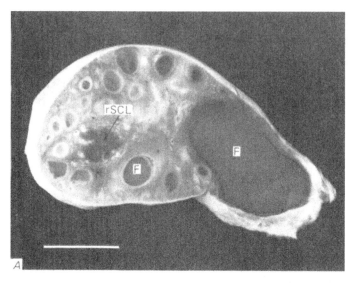

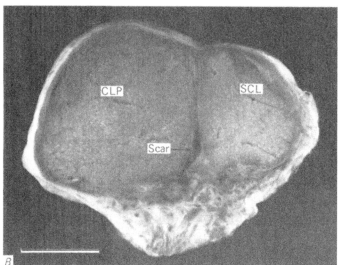

Figure 4.5. Examples of structures found in macroscopic sections of ovaries, with scale bar on all sections representing 5 mm. (A) Left ovary of 28-month-old in March, showing 1 mature follicle (11 mm), 1 follicle >2 mm and a regressing secondary CL (3 mm); other follicles present <2 mm diameter. (B) Left ovary of 54-month-old in July, showing a CL of pregnancy (10 mm diameter) and a secondary CL (8 mm); the dark structure sandwiched between the two CL is a distorted luteal scar. (C) Left ovary (above) of 60-month-old in December, showing a regressing secondary CL (3 mm) and a smaller, darker luteal scar, and right ovary (below) showing a regressing CL of pregnancy (5 mm); both ovaries have many follicles <2 mm diameter. (Photographs by C.J. Gilbert.)

parturition, the CL of pregnancy and any secondary CL degenerate over a period of months, and luteal scars formed from CL of pregnancy provide a record of past productivity of each animal (Figure 4.5B,C).

There are wide variations in the reproductive performance of *Rangifer* females, both in their age of first breeding and in their pregnancy rates (White *et al.*, 1981). The only previous descriptions of reproductive changes in female reindeer are a small series of ovarian weights (Roine, 1970), histological changes in ovaries (Borozdin, 1969) and measurements of seasonal differences in plasma progesterone concentrations (McEwan & Whitehead, 1980; Rehbinder, Edqvist, Riesten-Århed & Nordkvist, 1981; Ringberg & Aakvaag, 1982). In this section, I compare pregnancy rates of reindeer on South Georgia with other populations of *Rangifer*. Then I examine intraspecific differences in reproductive performance by comparing structures present in the ovaries of barren-ground caribou and South Georgia reindeer (McEwan, 1963; Dauphiné, 1976, 1978; Leader-Williams & Rosser, 1983).

Pregnancy rates were determined from shot reindeer, as described in Leader-Williams (1980b). Pregnancy rates were documented directly in females shot after mid-May when foetuses could be detected macroscopically. Pregnancy rates were documented indirectly in females shot from calving until February. This was achieved by examination of the reproductive tract, including ovaries, and by examination and weighing of the mammary gland. Pregnancy rates were not recorded from females shot between March and mid-May. Ovarian data are derived from a

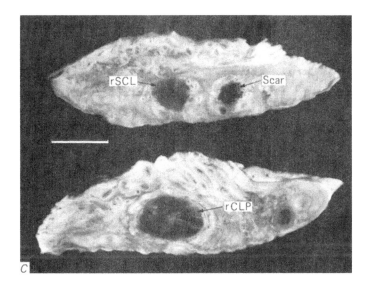

series of measurements of the following, as described in Leader-Williams & Rosser (1983): (1) ovary weight was measured for both sides; (2) examples of various ovarian structures are shown in Figure 4.5 and on the basis of Dauphiné's (1978) study, these were classified, measured and enumerated as one of (a) follicles of 2–5, >5–8 and ⩾8 mm diameter throughout the annual cycle, (b) corpora lutea (CL) of pregnancy and their accompanying secondary CL if present from April until November, (c) regressing CL from November until March, (d) luteal scars from previous pregnancies throughout the annual cycle, and (e) CL of ovulation in March and April; (3) certain ovaries were examined histologically to check for the presence of luteal structures.

4.31   Pregnancy rates

There were no signs of pregnancy in 33 female reindeer aged from 6 months (age of first possible conception) to 15 months (end of first possible lactation), examined from all herds on South Georgia. Furthermore, no yearling was accompanied by a calf or had a swollen udder (cf. Bergerud, 1964b) during herd composition counts. Thus reindeer calves on South Georgia do not conceive in their first autumn and give birth at 1 year of age. Pregnancy rates of reindeer over 18 months of age did not differ between either of the two winters of the study, between different herds, nor between assessments made by direct or indirect methods (Table 4.3). The combined results show that reindeer on South Georgia had pregnancy rates of over 90% during 1973–6. Furthermore, there are no differences in age-specific pregnancy rates, even for reindeer calving for the first time at 2 years of age (Figure 4.6).

Table 4.3. *Pregnancy rates of reindeer on South Georgia, assessed directly and indirectly (from Leader-Williams, 1980b)*

Herd	Age >1½ yrs, May–Nov. Total (N)	Pregnant (%)	Age >2 yrs, Nov.–Feb. Total (N)	Conceived (estimated) (%)	Age >1½ yrs, May–Feb. Total (N)	Pregnant or conceived (%)
Barff	80	90	34	91	114	90
Royal Bay			33	91	33	91
Busen	28	96	18	89	46	93
Total					193	91

Chi-square tests show no differences in pregnancy rates between the 1974 and 1975 winters ($\chi^2 = 0.55$, $P > 0.10$) nor between direct and indirect assessments ($\chi^2 = 0.04$, $P > 0.10$). There are no differences in pregnancy rates between reindeer calving at 2 years and at older ages ($\chi^2 = 0.70$, $P > 0.10$). Because of similarity in pregnancy rates, data from all herds are combined to examine age-specific pregnancy rates. This analysis shows no differences ($\chi^2 = 0.32$, $df = 3$, $P > 0.10$) in pregnancy rates even for combined age classes (reindeer calving at 2 and 3 years, 4 and 5 years, 6 and 7 years and ≥ 8 years).

Age-specific pregnancy rates vary in different reindeer and caribou sub-species and populations (Table 4.4). Amongst continental (as opposed to High Arctic island) sub-species, reindeer are generally able to achieve higher pregnancy rates at earlier ages than are caribou. Both wild and domestic reindeer calves can become pregnant under good nutritional conditions (Palmer, 1934; Skjenneberg & Slagsvold, 1968; Preobrazhenskii, 1969; Skuncke, 1969; Reimers, 1972, 1983b; Skogland, 1985), but the normal age of first calving is 2 years (Table 4.4). In contrast pregnancies have not been recorded in caribou calves, and their age at first calving can be as late as 3 or 4 years. Continental caribou may not achieve maximum pregnancy rates until 4 or 5 years of age, except in expanding populations (Parker, 1981). The cause of these differences is unclear, but presumably has a genetic basis (Bergerud, 1980a). The pregnancy rates of Peary caribou and Spitzbergen reindeer show much wider annual variations than those recorded in continental populations (Thomas, 1982; Tyler, 1987a). These arise from extreme fluctuations in climatic severity and physical condition of females in the High Arctic. A study in the High Arctic is still needed to determine whether infertility

Figure 4.6. Age-specific pregnancy rates of reindeer, using combined data from all herds. Sample sizes are shown above columns (data from Leader-Williams, 1980d).

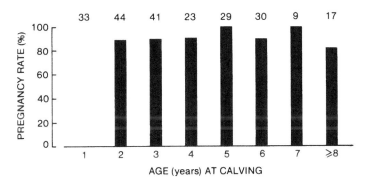

Table 4.4. *Age-specific pregnancy rates (%) amongst different populations of reindeer and caribou*

Sub-species or type	Location	Age of calving					Source
		1 yr	2 yrs	3 yrs	≥4 yrs		
Herded reindeer	Norway	6–18	52–85	70–78	70–87		Skjenneberg & Slagsvold (1968)
Wild reindeer	Norway – South	0–75	23–100	65–100	90–100		Reimers (1972, 1983b); Skogland (1985)
Tundra reindeer	USSR – Taimyr	0	80–90	—	—		Michurin (1967)
Herded reindeer	Sweden – Lapland	0	70	70–90	—		Skuncke (1969)
Spitzbergen reindeer	Svalbard	0	26–82	—	—		Tyler (1987a)
Introduced reindeer	South Georgia	0	89	90	92		This study
Woodland caribou	Canada – Labrador	0	11–43	22–90	95		Bergerud (1971); Parker (1981)
Alaskan caribou	Alaska	0	13	61	89		Skoog (1968)
Barren-ground caribou	Canada – NWT	0	2–33	48–50	80–90		McEwan (1963); Dauphiné (1976)
Peary caribou	Canada – Arctic islands	0	46	7–100	—		Thomas (1982)

arises from failure to ovulate, failure to come into oestrus or early loss of embryos. The remainder of this section concentrates upon a comparison between reindeer and barren-ground caribou.

4.32 Attainment of sexual maturity

The age at which females attain sexual maturity may be considered as the age when they first undergo oestrous cycles and are capable of becoming pregnant. Pregnancy on South Georgia first occurs in yearlings, but can occur in reindeer calves in the Northern Hemisphere under exceptional circumstances (Table 4.4). Examination of the ovaries of calves on South Georgia shows that ovarian weights and follicle numbers do not change until the second annual cycle, when pregnancies also first occur (Figure 4.7). Follicles over 5 mm appear to reach

Figure 4.7. Average ovary weights (mean ± SE shown only for 3 or more females per month class) and follicle numbers in 28 reindeer calves and 10 non-pregnant, nulliparous yearlings. Sample sizes are shown above follicle numbers (from Leader-Williams & Rosser, 1983).

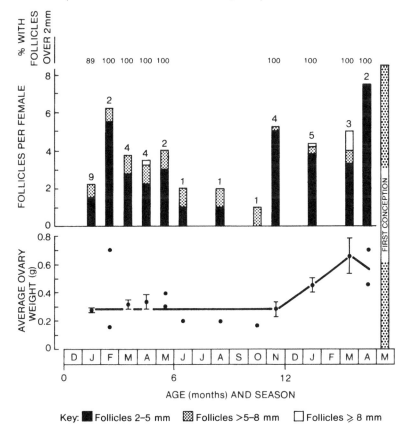

Key: ■ Follicles 2–5 mm   ▨ Follicles >5–8 mm   ☐ Follicles ⩾ 8 mm

Figure 4.8. The mean ± SE weights of ovaries bearing the CL of pregnancy and contralateral ovaries; follicle numbers; and mean ± SE diameter of CL of pregnancy and secondary CL in 136 breeding reindeer. Sample sizes are shown as appropriate (from Leader-Williams & Rosser, 1983).

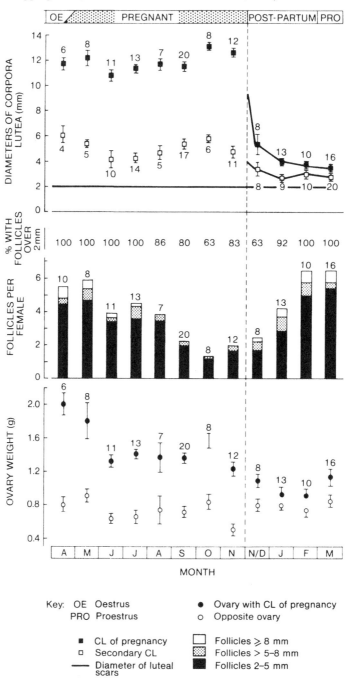

maturity, but the lack of luteal tissue indicates a failure of ovulation. From this lack of ovarian activity in calves, it may be concluded that physiological puberty does not occur on South Georgia until females become yearlings. Examination of barren-ground caribou ovaries shows that no calves, only 57% of yearlings and 80% of 2-year-olds had ovulated (Dauphiné, 1976). Therefore, the onset of ovarian activity coincides with the age of first breeding in barren-ground caribou and reindeer. Lack of pregnancies in both immature reindeer and caribou arises from a lack of ovarian activity rather than from embryonic loss. However, the onset of ovarian activity and of puberty in females varies considerably between the sub-species of *Rangifer* so far studied.

Analysis of variance shows that ovarian weights of calves from each month do not differ ($F_{5,42} = 1.82$, $P > 0.10$), but those of yearlings increase from 12 to 16 months of age ($F_{3,24} = 5.11$, $P < 0.01$). The numbers of follicles >2 mm were ranked and compared by the Kruskal-Wallis one way analysis of variance, which shows that follicle numbers in calves do not change ($H = 10.36$, $df = 5$, $P > 0.05$).

4.33   *Ovarian changes in yearling and adult females*

More than 90% of female reindeer over 16 months of age on South Georgia can be classed as breeding. That is to say they either were pregnant or had given birth during the annual cycle of their collection. At all stages of the annual cycle, reindeer have from two to three times as many follicles as caribou (Figures 4.7, 4.8, cf. Dauphiné, 1978). Furthermore, reindeer have twice as many secondary CL as caribou (Figure 4.9). The size of our sample at the time of ovulation did not allow a direct estimate of ovulation rate (cf. Dauphiné, 1978). However, the greater frequency of CL of pregnancy and differences in numbers of secondary CL (Figure 4.9) suggest a higher rate of ovulation in reindeer. This raises the question of whether their ovaries are more productive than those of caribou.

There is no difference between the numbers of secondary CL in the ovary with the CL of pregnancy (35) compared with the number in the opposite ovary (37). Chi-square test shows that the frequencies of secondary CL do not change ($\chi^2 = 4.73$, $df = 7$, $P > 0.10$) throughout the different months of pregnancy. Reindeer have more secondary CL in their ovaries than do barren-ground caribou ($\chi^2 = 40.20$, $P < 0.001$).

4.34   *Ovarian productivity*

Records of previous CL of pregnancy are preserved as luteal scars (Figure 4.5), which accumulate and add to ovarian weight in successive years (Figure 4.10). The CL of the current pregnancy occurs as frequently in the left as in the right ovary of both reindeer and caribou. However, individual caribou had a dominant ovary which produced the majority of CL of pregnancy during their lifespan (Dauphiné, 1978). In contrast, the ovaries of individual reindeer are equally productive. Both

Figure 4.9. Numbers of secondary corpora lutea in ovaries of reindeer in different months, and of barren-ground caribou for all months of pregnancy combined (data from Leader-Williams & Rosser, 1983; Dauphiné, 1978).

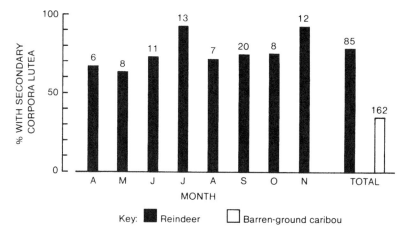

Figure 4.10. Regression of weights of ovaries bearing the CL of pregnancy (solid circle) and of the contralateral ovary (open circle) on age of 72 pregnant reindeer: sample sizes and respective regression equations are shown (from Leader-Williams & Rosser, 1983).

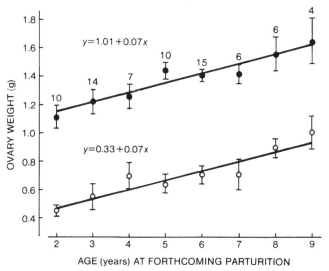

the ovary bearing the CL of pregnancy and the other ovary show a parallel increase in weight with age in reindeer (Figure 4.10), the CL of pregnancy did not occur more frequently in the ovary of the pair with the most scars. The reasons for, or the consequences of, this difference between reindeer and caribou are not apparent.

In pregnant reindeer, the CL of pregnancy occurred 43 times on the left, and 42 times on the right side; the respective figures for regressing CL of pregnancy were 33 and 24. The ratio of all CL of pregnancy did not differ from unity ($\chi^2 = 0.38$, $P > 0.10$) between left and right sides. In 87 pregnant and post-partum reindeer >3 years of age at their last parturition, the functional or regressing CL of pregnancy occurred in 36 ovaries with fewer scars than the contralateral ovary, and in 37 ovaries with more scars; in the remaining 14 pairs of ovaries, luteal scars occurred in equal numbers. Thus the CL of pregnancy did not occur more frequently ($\chi^2 = 0.98$, $P > 0.10$) in the ovary of the pair with most scars.

Counts of numbers of luteal scars have been of value in determining the productivity of several species of cervid, including caribou (Simkin, 1965; Thomas, 1970; Mansell, 1971; Dauphiné, 1978). However, in some studies persistence of CL of ovulation and of secondary CL has resulted in inaccurate estimates (Golley, 1957; Morrison, 1960). No luteal scars of regressing CL of pregnancy are found in female reindeer which were calves at the previous calving season. The regression of luteal scars on age is significant and the slope (= pregnancy rate) of 0.93 is in close agreement with the value of 0.90 derived from examination of uteri (Table 4.3). Furthermore, the regression reflects accurately the higher reproductive performance of reindeer compared with barren-ground caribou, both in differences of their ages of first breeding and of their pregnancy rates (Figure 4.11).

In reindeer, CL of pregnancy and secondary CL decrease in size after parturition, but both structures are distinguishable from each other by their different sizes (Figures 4.5, 4.8). During the post-partum period both structures are also distinguishable from luteal scars of previous pregnancies by their larger size, lighter colour, and rounder shapes than the more distorted luteal scars, which are invariably less than 2 mm in diameter. However, the regressing secondary CL become macroscopically indistinguishable from luteal scars after April, and technical difficulties described in more detail elsewhere did not allow us to distinguish histologically between regressed secondary CL and luteal scars (Leader-Williams & Rosser, 1983). Thus reindeer differ from caribou in which secondary CL regress quickly and absolutely, and as a result are not included erroneously in counts of luteal scars (Dauphiné, 1978). We made counts of luteal scars in both ovaries of all reindeer over 1 year of age. In those collected in the post-partum period, the count was increased by 1 when a regressing CL of pregnancy was present, to include the recent pregnancy. In our analysis of counts of luteal scars to assess productivity, we proceeded with caution and bore in mind the probability of having included extra scars which resulted from secondary CL (Leader-Williams & Rosser, 1983).

Figure 4.11. (A) Regression of the numbers of luteal scars (solid bar, mean ± SE on the age of 155 reindeer over 1 year of age ($r = 0.87$, $P < 0.001$), shown with sample sizes and regression equation; (B) age-specific fecundity rates (= 0.5 × pregnancy rates) of reindeer (solid line). Comparative data for caribou are shown with a broken line (data from: Dauphiné, 1976, 1978; Leader-Williams, 1980b; Leader-Williams & Rosser, 1983).

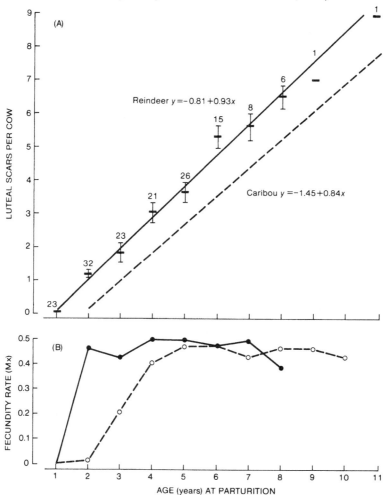

4.35 Causes of reproductive failure

Reindeer sustain a high pregnancy rate throughout life, but the small sample of non-breeding reindeer on South Georgia was examined for the cause of their reproductive failure (Leader-Williams & Rosser, 1983). In caribou, non-ovulation was an important cause of infertility in young animals, whereas non-conception after ovulation was the usual cause in older females (Dauphiné, 1976). In contrast, examination of non-pregnant reindeer showed that ovulation had occurred in several young and all older females (Figure 4.12). This infertility in young reindeer mainly arises from non-conception after ovulation. The differences documented here suggest further work to investigate the ovarian characteristics of *Rangifer* would be valuable, most especially for High Arctic sub-species in which rates of reproductive failure vary annually (Thomas, 1982; Tyler, 1987a).

Reindeer from all herds that were not pregnant or had not given birth in their year of collection were examined for their cause of infertility. Data for these reindeer are listed in detail in Leader-Williams & Rosser (1983). For the present analysis, reindeer were divided according to their age and reproductive status and according to whether they had ovulated or had lost their conceptus. Fisher's exact probability test shows that there is a difference ($P = 0.05$) between the numbers of young and older reindeer that conceived.

Figure 4.12. Number of young and older non-breeding reindeer that failed or succeeded to ovulate and to conceive (data from Leader-Williams & Rosser, 1983).

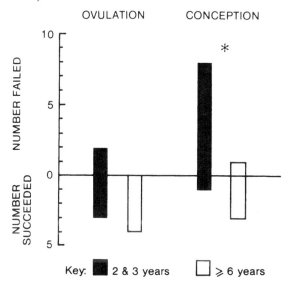

4.4   Reproductive changes in the male

The testis has two distinct functional components (Figure 4.13).
The seminiferous tubules undergo cycles of spermatogenic activity to
produce spermatids which in turn develop into spermatozoa. After a

Figure 4.13. Testis section (6 μm) stained in haematoxylin and eosin, from
16-month-old reindeer at the rut. Scale bar represents 50 μm. Seminiferous
tubules have spermatids near their lumina and interstitial tissue is wedged
between tubules. (Photograph by C.J. Gilbert.)

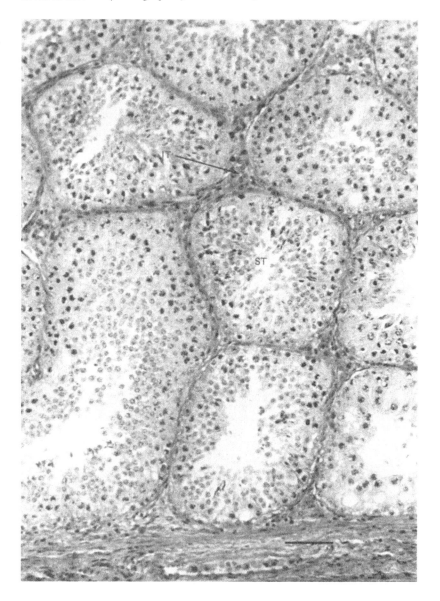

period of maturation in the epididymis, spermatozoa are ejaculated at copulation. The interstitial tissue, by contrast, produces male sex hormones, collectively known as androgens, which are largely responsible for the onset of male sexual behaviour and the development of secondary sexual characters. In this section, I consider in turn the reproductive changes leading to the onset of sexual maturity, and those occurring in sexually mature male reindeer and caribou.

Data are derived from a series of measurements of the following, as described in Leader-Williams (1979b): (1) combined testis and epididymis weights were measured for both sides, but as the variation for the remainder was only 8±2.2%, data are presented only for the left side; (2) seminiferous tubular tissue area was measured in 20 tubules cut transversely in histological sections of each testis; an index of spermatogenesis was derived from the numbers of tubules out of 20 containing spermatids, and an index of 5 or over was considered to indicate full spermatogenesis; (3) the period of fertility was assessed by scoring the cauda epididymis for the presence or absence of spermatozoa, and for cellular debris including immature cells which indicate the end of this period; (4) plasma testosterone concentrations were assayed using the specific radioimmunoassay of Corker & Davidson (1978). A small sample of males had abnormal testes (Leader-Williams, 1979c) and will not be considered here.

4.41 *Attainment of sexual maturity*

Physiological puberty in male ungulates has been defined as the period between the initial onset of androgenesis and the first appearance of spermatozoa in the testis (Davies, Mann & Rowson, 1957; Lincoln, 1971a). Previous work on reindeer and caribou had not clearly defined when puberty occurred in males. There is a cycle of spermatogenic activity in barren-ground caribou calves, but puberty was said not to occur until 17 to 18 months of age, on the first appearance of spermatids in the testis (McEwan, 1963). However, a rise in the concentration of plasma testosterone was recorded at 4 to 5 months of age in two captive hybrid reindeer and caribou calves (Whitehead & McEwan, 1973).

To clarify changes leading up to puberty, I examined reindeer calves from birth to 12 months of age (Leader-Williams, 1979b). Their testis and epididymis weights increase dramatically during the first few months of life (Figure 4.14). A similar increase in the ratio of testis and epididymis weight to carcass weight (Figure 4.15) shows clearly that increases in testis and epididymis weight cannot be attributed solely to

Figure 4.14. Variations in mean ± SE weight of left testis and epididymis; seminiferous tubular tissue area; plasma testosterone concentrations; and period of full spermatogenesis in 107 male reindeer. Sample sizes are shown as appropriate. The stippled vertical bars represent the rut (redrawn from Leader-Williams, 1979b).

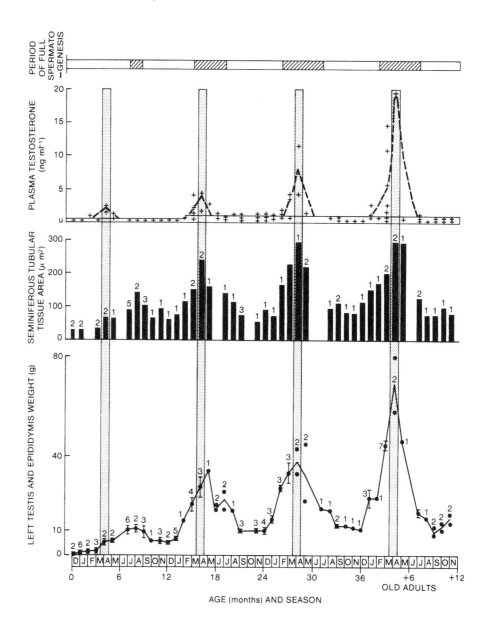

rapid growth of calves during their first summer (Chapter 6). Testis weights of calves are heaviest at 7 to 9 months of age, several months after the rut (Figure 4.14).

> Analysis of variance shows that testis and epididymis weights from each month differ ($F_{8,21} = 14.10$, $P < 0.001$), as do ratios of testis and epididymis weight to carcass weight ($F_{8,21} = 13.55$, $P < 0.001$). Student's *t* test between six month classes shows that neither the weight nor the ratio differ between 7 and 8 to 9 months ($t = 0.28$ and 0.37, $df = 9$, $P > 0.10$), but that they do differ between other month classes (Figure 4.15).

Spermatogenic activity is also evident in the testes of calves. Seminiferous tubules show signs of histological change and the tubular tissue area increases from 4 to 5 months of age and is at its greatest at 8 months (Figure 4.14). Spermatids first appear in the seminiferous tubules at 8 months, and spermatozoa may appear in the epididymis of 9-month-old males (Figure 4.16). The efficiency of spermatogenesis decreases by 10 months and the tubules become re-organised for the subsequent cycle by 12 months. These results amplify those for caribou calves (McEwan, 1963). There is also a cycle of androgenic activity in reindeer calves (Figure 4.14). Testosterone levels in plasma reach 2.3 ng ml^{-1} at 4 months

Figure 4.15. Changes in the mean left testis and epididymis weight (g) of reindeer calves related to age and carcass weight (kg) (data from Leader-Williams, 1979b).

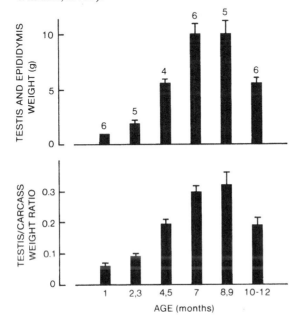

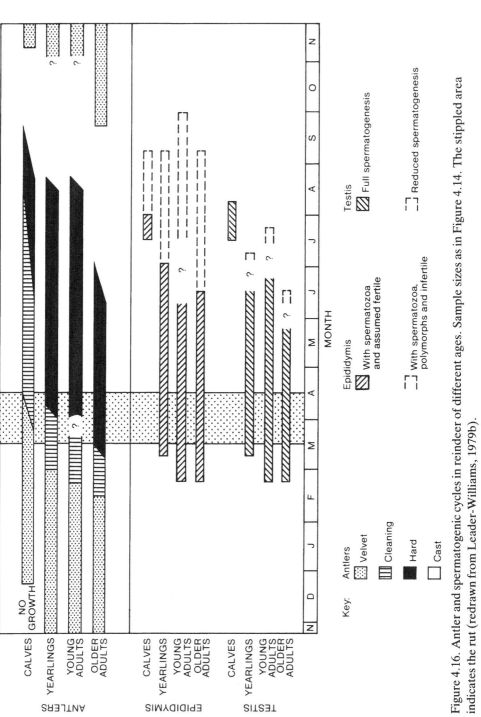

Figure 4.16. Antler and spermatogenic cycles in reindeer of different ages. Sample sizes as in Figure 4.14. The stippled area indicates the rut (redrawn from Leader-Williams, 1979b).

of age and fall to basal levels by 6 months, in agreement with timing of the androgenic cycle noted by Whitehead & McEwan (1973). From evidence of this androgenic cycle and of spermatogenic activity, it may be concluded that male reindeer on South Georgia attain physiological puberty at 4 to 8 months of age. Furthermore, it appears that puberty occurs consistently at 4 to 8 months in male *Rangifer* (McEwan, 1963; Whitehead & McEwan, 1973).

Even though puberty may occur in calves, it is of little direct reproductive value at that time. Androgenesis coincides with time of the rut, but spermatozoa are not found until three months afterwards. Furthermore, rutting behaviour, known largely to be under the control of testosterone in red deer (Lincoln, Youngson & Short, 1970; Fletcher, 1978), is not seen amongst *Rangifer* calves on South Georgia or elsewhere (Espmark, 1964a; Lent, 1965a; Henshaw, 1970; Bergerud, 1973, 1974a; Thomson, 1977), though play behaviour is observed (Espmark, 1971a). With the exception of antlers (Section 4.5), no secondary sexual characters like swollen neck girth, neck mane or rutting odour (Lincoln, 1971b) develop in reindeer calves. However, male deer require several months to initiate spermatogenesis at puberty (Lincoln, 1971a). The indirect reproductive consequence of physiological puberty in *Rangifer* calves may be to ensure that the testis is fully functional in yearlings at the rut. Early puberty will also have a cost, and is likely to be a factor that enhances differential mortality between the sexes (Chapter 7).

4.42   *Seasonal cycle in yearling and adult males*

Seasonal changes in testis weight and histology and in plasma testosterone of *Rangifer* (Meschaks & Nordkvist, 1962; McEwan, 1963; Meschaks, 1966; Whitehead & McEwan, 1973; Whitehead & West, 1977; Ringberg Lund-Larsen, 1977; Mossing & Damber, 1981) are similar to those of other temperate cervids (e.g. Lincoln, 1971b). However, the large sample of reindeer collected on South Georgia provided information on age-related changes (Leader-Williams, 1979b). Reindeer over 1 year were divided into three age classes: yearlings (13 to 24 months); young adults (25 to 36 months); older adults (37 months and older). Minimum testis and epididymis weights are similar in each year class. However, there is an amplitudinal increase in maximum weights with age, the increases being *c*.2-, 3 to 4- and 4-fold in yearlings, young adults and older adults respectively (Figure 4.14).

> The minimum testis and epididymis weights of yearlings collected between 15 July and 15 December are heavier than those of calves (*t* =

4.56, $df = 15$, $P < 0.001$), showing that further growth of the testis and epididymis takes place after the first testicular cycle. However, analysis of variance shows that the minimum weights of 10.9 ± 0.8 g ($N = 11$), 13.0 ± 1.3 g ($N = 5$) and 12.0 ± 1.0 g ($N = 8$) for yearlings, young and old adults did not differ ($F_{2,21} = 1.09$, $P > 0.10$). By contrast, the maximum testis weights from reindeer collected between 15 February and 15 May of 24.6 ± 2.6 ($N = 7$), 34.6 ± 3.2 g ($N = 7$) and 49.7 ± 3.8 g ($N = 10$) all differ ($F_{2,21} = 13.94$, $P < 0.001$). The regression of the maximum testis and epididymis weight on age for these 24 reindeer was significant ($r = 0.691$, $P < 0.001$) and of the form $y = 13.1 + 7.02x$.

The production of spermatids begins, and spermatozoa appear in the epididymis about one month before the rut among males of 2 years and older. However, spermatid production coincides with the start of the rut in yearlings. The production of spermatids ceases at the end of June and spermatozoa, mostly degenerative and few in number, are present in the epididymis with polymorphs and cellular debris until September in adults and November in yearlings (Figure 4.16). Peak values of seminiferous tubular tissue area occur around the time of the rut in adults and yearlings (Figure 4.14). However the amplitude of this cycle does not increase markedly with age, the increase being about 2.5-, 3- and 3-fold in each age class. This contrasts with the cyclical rise and fall in plasma testosterone concentrations, in which there is a marked increase in the amplitude of the cycle with increasing age (Figure 4.14).

In fallow deer a progressive increase in peak testis and epididymis weights was found throughout all age classes. The first increase from calf to yearling was associated with the onset of puberty in yearlings, but no explanation was given for the continued increases in weight with age (Chaplin & White, 1972; Chapman & Chapman, 1975). In reindeer it appears that much of the testicular weight increase with age arises from increased secretory activity, for it is plasma testosterone rather than seminiferous tubular tissue area that shows a parallel increase with age (Figure 4.14). Whilst measurements of testosterone concentrations in single plasma samples do not quantify rapid diurnal variations arising from episodic secretion (Lincoln & Kay, 1979; Stokkan, Hove & Carr, 1980), it is clear that the onset and duration of testosterone secretion by the testis is rapid and short in reindeer (Figure 4.14; Whitehead & West, 1977; Ringberg Lund-Larsen, 1977). Furthermore, it follows that the relative rate of increase and of decrease in testosterone secretion in relation to time changes with age, as suggested for red deer (Lincoln, 1971a). Such changes might explain many facets of male sexual behaviour and physiology. For example, the age-related shifts seen in the onset of sperm production (Figure 4.16), the earlier onset of sexual behaviour in older reindeer males and their more dominant position in the hierarchy at this time (Espmark, 1964a; Bergerud, 1973, 1974a), and certain events in the antler cycles of males.

4.5   The antler cycle

Reindeer and caribou are unique amongst cervids because females regularly bear antlers. It has been known since the time of Aristotle that antler growth in males is related to activity of the testis, as confirmed since for several species (Chapman, 1975; Goss, 1983). Experimental castration has shown that the cycle of antler growth and shedding is largely under the control of testosterone (Wislocki, Aub & Waldo, 1947). Therefore, antlers are regarded rightly as male secondary sexual characters.

But what is the situation in *Rangifer*? Not only do females carry antlers, but early work suggested that events of the antler cycles of both sexes are largely independent of gonadal activity (Tandler & Grosz, 1913). Antler cycles of male and female *Rangifer* also have other unique features (Figure 4.17). For example, one antler cycle is completed during the first year of life, and in males there is a delay of up to 4 months between casting of the old antlers after the rut and regrowth of the new antlers in the spring (Leader-Williams, 1979b). In this section, I describe the antler cycles of reindeer on South Georgia, and discuss possible hormones that are involved in its control.

> Antlers were classified as follows: never grown; growing in velvet; cleaning the velvet; partly clean (having desiccated tatters of velvet at the tips); hard; cast. Examples of the different stages of the antler cycle are shown in Figure 4.17 and others. Plasma testosterone concentrations were assayed in females using the specific radioimmunoassay of Corker & Davidson (1978).

4.51   *Timing of events*

The timing of the antler cycles in male and female reindeer on South Georgia is summarised in Figure 4.18. Development of antler pedicles is not observed in reindeer foetuses either on South Georgia or in Norway (Wika, 1980, 1982), though patches of dark hair occur over the future pedicle area after the first third of pregnancy. Hence, reindeer differ from red deer, in which there is a transitory enlargement of the pedicle in early foetal life as a result of the testosterone surge following

sexual differentiation (Lincoln, 1979). At birth the hair over the pedicle area is dense, black and well circumscribed (Figure 3.8A). Growth of the antler bud begins in December in both sexes (Figure 4.17A), and macroscopic differentiation of the antler and the pedicle becomes clear only after the antler is longer than 5 cm (Figure 4.19). Again reindeer differ from other species in which pedicle formation occurs several months before the first antler starts to grow (Goss, Severinghaus & Free, 1964; Lincoln, 1971a).

The timing of events early in the first antler cycle is similar in both sexes. Antlers grow in velvet until March or April, and the process of cleaning the velvet is prolonged and incomplete (as in Figure 4.17D). Casting of antlers in male calves precedes that in females by a few weeks, and males have a short period without antlers before regrowth of their second set. A major divergence occurs between the sexes in yearlings and adults (Figure 4.18). In males, antler cleaning starts progressively earlier and is more thorough and rapid. Older males spend progressively longer with cast antlers (Figure 4.18; Meschaks & Nordkvist, 1962; Whitehead & McEwan, 1973; Bergerud, 1976), and the pedicle remains in a refractory state for periods of up to 4 months. In females, the only shift seen is in the time of cleaning in calves and yearlings, and thereafter the timing of the cycles with age does not change (Figure 4.18). In addition, many females remain with only partially clean antlers until casting

Figure 4.17. The antler cycle of reindeer has several points of interest. Male and female calves complete an antler cycle in their first year of life (A and D); adult males shed their antlers after the rut (B), whereas female antlers are retained throughout the winter (C) and are shed at around the same time as those of calves in spring.

Figure 4.18. Antler cycles of males and females. Sample sizes are shown as appropriate.

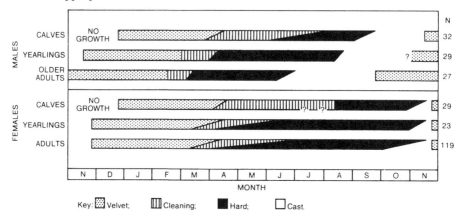

Figure 4.19. Antler growth preceeds the macroscopic differentiation of the pedicle. A, 1 to 2 months of age; B, 2 to 3 months; C, >4 months.

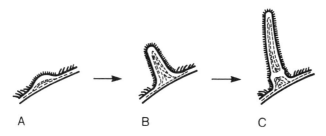

Figure 4.20. Numbers of pregnant reindeer that had cast one or two antlers before giving birth. The difference between samples in September, October and November differed ($\chi^2 = 8.00$, $df = 2$, $P < 0.025$), but those for October and November did not ($P > 0.10$). The percentages with cast antlers are shown with dotted lines, together with sample sizes for each period.

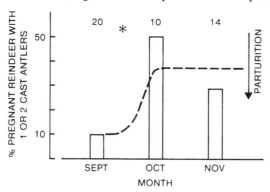

(Figure 4.17C). Casting takes place in yearlings and adult females around the time of calving and regrowth of the new antlers commences almost immediately. A relationship between the events of calving and of antler casting has often been suggested because barren females cast one month before pregnant females, which do so only after giving birth (Kelsall, 1968; Lent, 1965b; Espmark, 1971b; Bergerud, 1976). My sample of non-pregnant females was small and proved nothing. However, several pregnant females had cast before calving on South Georgia (Figure 4.20), and similar observations have been made in Greenland (Thing, Olesen & Aastrop, 1986). Therefore ratio counts of antlered to unantlered females around the calving season used as a method of assessing either timing of the calving season or calving percentages should be interpreted with care.

4.52   *Hormonal control*

The antler cycles of reindeer and caribou provide an as yet unanswered physiological paradox. Surgical removal of the testis and ovary is said to have less dramatic effects on the antler cycles than castration in other species of cervid (Tandler & Grosz, 1913; Hadwen & Palmer, 1922). This led to suggestions that sex hormones play an unimportant role in the antler cycles of *Rangifer* (Bubenik, 1975a; Goss, 1983). Alternatively, it was suggested that female reindeer might have high levels of circulating androgens in their blood (Henshaw, 1968a). Indeed, the abnormal occurrence of antlers in other female deer is often associated with testosterone-secreting tumours (reviewed by Chapman, 1975).

Contrary to Henshaw's suggestion, plasma testosterone concentrations in calves and adult females, assayed in the same way as those of males (Figure 4.14), never became elevated above 0.3 ng ml^{-1} ($N = 28$). Similarly, there was no correlation between the cycle of progesterone secretion and the antler cycle (McEwan & Whitehead, 1980). Evidence from castrations suggests that absence of gonadal hormones only modifies rather than prevents the completion of an antler cycle. Therefore it seems probable that a non-gonadal hormone which is common to both males and females is primarily involved. A likely candidate is prolactin (Wislocki *et al.*, 1947), a hormone that has been the focus of research in other areas of *Rangifer* biology (Ryg & Jacobsen, 1982a,b,c). The hormone(s) that are finally invoked will need to account for simultaneous growth of the first antler and development of the pedicle (Figure 4.19), and the other differences seen between the antler cycles of *Rangifer* and other species (Goss, 1983).

But is testosterone not involved at all in the antler cycle of males? The obvious sex differences between the antler cycles and evidence from castrations (Hadwen & Palmer, 1922) shows that it is. The first antler cycle of males (and all those of females) resemble the first cycle in red deer, in which cleaning is a lengthy process and continues long after the influence of the testis has waned (Lincoln, 1971a). The antler growth of yearling and adult male reindeer begins at or after the spring equinox (Figure 4.16) when testis weights and testosterone levels are minimal. Cleaning occurs as testis weights and testosterone concentrations are rising, and casting occurs subsequent to their decrease. Furthermore, the parallel age shift in the onset of production of spermatids and in antler cleaning suggests that both processes may occur under the influence of the same hormone. Therefore, these observations are consistent with the hypothesis that the presence of testosterone is important for the normal antler cycle of male *Rangifer*. Furthermore, an interspecific comparison of *Rangifer* with other temperate cervids shows an inverse relationship between length of the period spent without antlers and the length of the spermatogenic cycle (Figure 4.21). I suggest that testosterone overrides whichever primary hormone controls the antler cycle of both sexes. As a result, males are prepared in advance of the rut with clean antlers, to enhance their fighting ability and reproductive success.

Figure 4.21. Comparison of periods of full spermatogenesis in males of temperate species of deer (A) with the period in which males have cast antlers and (B) with the length of the calving season. 1, *Capreolus capreolus*; 2, *Odocoileus virginianus*; 3, *O. hemionus*; 4, *Alces alces*; 5, *Dama dama*; 6, *Rangifer tarandus*; 7, *Cervus elaphus*; 8, *C. e. canadensis*. (Data for (A) from: Short & Mann, 1966; Wislocki, 1943 and Mirachi *et al.*, 1977; West & Nordan, 1975; Chaplin & White, 1972 and Chapman & Chapman, 1975; this study; Lincoln, 1971b; Flook, 1970 and McCullough, 1969. Data lacking for *A. alces*. Data for (B), shown as period when 80% of calves are born with annual variations if available, from: Strandgaard, 1972a; Ransom, 1966; Markgren, 1969; Rieck, 1955; Nowasad, 1975; Guinness *et al.*, 1978a; McCullough, 1969.)

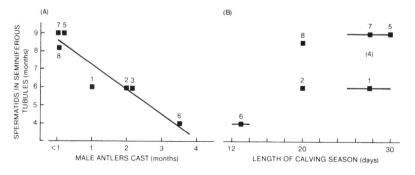

4.6   Reproductive potential and sexual dimorphism

Reindeer and caribou are a recent addition to the deer family, and have evolved to occupy very high latitudes and a far more open habitat than most other temperate species (Table 1.2). *Rangifer* show both differences and similarities in their reproductive characteristics to these other species, some of which I will discuss in this section.

Reindeer and caribou have short and highly synchronised mating and calving seasons (e.g. Dauphiné & McClure, 1974; Bergerud, 1975), the latter of which is about half the length of other temperate species of deer (Figure 4.21). They calve usually on the open tundra, often when migrating, and with predators in attendance (Kelsall, 1968; Parker, 1972; Semenov-Tian-Shanskii, 1975). As with migratory wildebeest in Africa (Estes, 1966; Watson, 1969) the defence of calves from predation is achieved in the safety of numbers, both by synchrony of births and by the formation of large groups after calving (Pruitt, 1960a; Lent, 1966; Bergerud, 1974b; Thomson, 1977; Figure 4.1). Calves do not remain hidden to escape from predators, unlike all other species of cervid, but instead follow their mothers from within one hour of birth (Lent, 1974). This may explain why *Rangifer* are the only species without a dappled coat at birth, which in other deer acts as camouflage for calves that remain hidden to avoid predators (Lent, 1974).

The reproductive potential of temperate cervids differs and species fall into two clear groups (Figure 4.22). Females of forest and woodland dwelling species frequently bear twins and tend to live in smaller groups,

Figure 4.22. Comparison of (A) maximum pregnancy rates and (B) testis weights in temperate species of deer forming smaller and large breeding groups. Species as in Figure 4.17. (Data for (A) from: Andersen, 1953; Cheatum & Severinghaus, 1950; Taber & Dasmann, 1957; Markgren, 1974; Chapman & Chapman, 1975; this study; Mitchell, 1973; Flook, 1970. Figure (B) from Clutton-Brock *et al.*, 1982.)

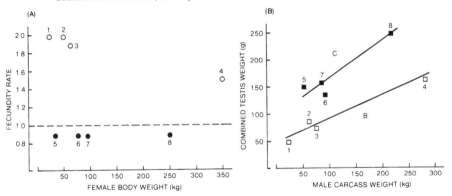

whereas those living in more open areas, including *Rangifer*, rarely have twins and live in larger groups (groups B vs. C of Clutton-Brock *et al.*, 1982). It is generally accepted that the female holds the key to the male's reproductive strategy (Trivers, 1972; Short, 1980), so what are the consequences of these differences? In group B, males have smaller antlers (Figure 1.2) and testes (Figure 4.22) relative to their body size and command access to only a small number of breeding females. However, they have a higher reproductive success per female mated providing each is successful in rearing her twins. In group C, the situation is reversed, and to achieve a high reproductive success males have to command access to a large number of females. Thus polygyny is well developed, antlers and testis size are large, and sexual dimorphisms in many aspects of their physiology and ecology accentuated. For reindeer and caribou in particular, the early and consistent occurrence of puberty in male calves, the short breeding season (in calving, oestrous and spermatogenic activity) and the intense rut centred around them, are of additional importance compared with other temperate species of deer.

4.7   Summary

1. Reindeer have reversed their breeding season by 6 months on South Georgia, and rut in March and April, and calve in November. No drift in the reversal has occurred compared with Norwegian reindeer.
2. Reindeer females on South Georgia first conceive as yearlings and have high pregnancy rates of >90%. There is wide variation in pregnancy rates amongst reindeer and caribou.
3. Reindeer females on South Georgia do not attain physiological puberty as calves. Compared with barren-ground caribou, yearling and adult reindeer show an earlier onset of ovarian activity, have twice as many follicles >2 mm diameter and secondary corpora lutea. These differences reflect the higher reproductive rate of reindeer.
4. Reindeer males achieve physiological puberty as calves. The amplitude of the cyclical change in testis weight and plasma testosterone increase with age, and can be correlated with the earlier onset of events in the spermatogenic cycles of other animals.

5. The antler cycles of males and females differ in their timing. Testosterone has an important influence on the antler cycle of males. However, a non-gonadal hormone common to males and females probably has primary control over the antler cycle of both sexes.

6. In adapting to life at high latitudes, reindeer and caribou have highly synchronised calving seasons and short spermatogenic cycles compared with other temperate species of deer. In common with other temperate species which are highly polygynous and form large breeding groups, reindeer and caribou have heavy testis weights and only bear single young.

5

Food habits

It is astonishing how [the reindeer] can get at his proper food through the deep snow that covers it, and by which it is protected from severe frosts.
Carl Linnaeus (1732)

5.1 Introduction

The success of any ungulate living in a seasonal environment depends on its ability to find and eat forage of sufficient quality and quantity to stay alive and to reproduce. Temperate cervids live in habitats where food quantity is limited in winter, and where food quality is at its best in spring and summer (Klein, 1970a). In their native habitats of the far north, reindeer and caribou feed in open areas during summer and select nutritious herbage, mainly grasses, forbs, sedges and shrubs, during the relatively short, snow-free growth period (e.g. Skogland, 1980, 1984; White *et al.*, 1981). In autumn many continental reindeer and caribou migrate to woodland areas where they overwinter and forage for lichens and smaller amounts of various vascular species (e.g. Aleksandrova, 1937; Klein, 1980a, 1986).

The flora on South Georgia differs markedly from that encountered by reindeer and caribou in the Northern Hemisphere. It is both poor in species, especially of vascular plants, lacks true shrubs and had never been grazed by any mammalian herbivores. Furthermore, seasonal migrations could not be undertaken in the restricted ranges available between glacier boundaries. However, South Georgia had some similarities to reindeer habitats in the Northern Hemisphere. Certain plants, notably 'reindeer lichens' of the genus *Cladonia*, have a bipolar distribution, and snow-fall restricts the availability of forage during winter.

In dimorphic cervids, such as *Rangifer*, the nutritional requirements of males and females are likely to differ (Clutton-Brock *et al.*, 1982). Therefore, I aimed to determine whether male and female reindeer selected diets of different composition and quality on South Georgia. It was also of interest to determine which plants are selected because of the effect of reindeer on species composition in plant communities since their introduction (Bonner, 1958; Lindsay, 1973; Kightley & Lewis Smith, 1976). I confine this chapter to a discussion of forage availability and selection during my study, and later speculate how changes in vegetation composition arose (Chapter 9). First I describe the forage available to reindeer during snow-free periods (5.2) and how snow limits the use of plant communities on South Georgia (5.3). Then I describe seasonal changes in species composition of the diet, some possible factors that influence selection of forage plants (5.4), and seasonal changes in forage consumption of an important macronutrient (5.5). Finally, I discuss differences in the feeding ecology of *Rangifer* and other temperate cervids, and the differences between *Rangifer* in the Northern and Southern Hemispheres (5.6).

5.2 Forage availability during snow-free periods

Lewis Smith & Walton (1975) have defined eight different plant communities on South Georgia in an area close to Grytviken and not occupied by reindeer. These communities arranged according to altitudinal gradient comprise:

1. Tussock grassland: the climax vegetation of the coastal lowlands and the most productive community (Table 5.1). It is dominated by the winter-green coastal tussock grass *Poa* (= *Parodiochloa*) *flabellata*, which may grow up to 2 m in height on wet raised beaches near the shore. On drier slopes tussock grass tends to be shorter and is interspersed with mossbanks.

2. Mesic meadow: dominated by the hair grass *Deschampsia antarctica*.

3. Mossbank: dominated by *Chorisodontium aciphyllum* and *Polytrichum alpestre* with macrolichens.

4a. Dry meadow: dominated by the short tussock forming grass *Festuca contracta*, with the burnet *Acaena magellanica* and macrolichens in more exposed situations.

4b. Dwarf-shrub sward: dominated by *A. magellanica*, usually with a dense understorey of the moss *Tortula robusta*, on slopes and stream terraces.

5. Oligotrophic mire: dominated by the rush *Rostkovia magellanica* and by bryophytes.

6. Eutrophic mire: dominated by the rushes *Juncus scheuchzerioides* and *R. magellanica* and by bryophytes, especially *T. robusta*.

7. Fellfield: dominated by the grass *Phleum alpinum* and by cryptogams.

I have modified this description of communities slightly (Figure 5.1) to take account of the effects of reindeer grazing (Bonner, 1958; Lindsay, 1973; Kightley & Lewis Smith, 1976). Extensive stands of tussock grassland have been overgrazed (see Figures 9.1, 9.4). In some localised

Table 5.1. *Peak above-ground standing crop and production data for ungrazed plant communities on South Georgia (data from Lewis Smith, 1984)*

Community	Standing crop (biomass) $(g\, m^{-2})$	Net annual production $(g\, m^{-2}\, y^{-1})$
Tussock grassland[a]	15 000–16 000	10 000
Mesic meadow		300–600
Dry meadow	310–590	180–290
Oligotrophic mire		190–460

[a] Data are shown for individual mature plants in beach areas; data for the community are up to one quarter of the above values (Lewis Smith, 1984).

Figure 5.1. Vegetation types and major species on a typical reindeer area of South Georgia (from Leader-Williams, Scott & Pratt, 1981).

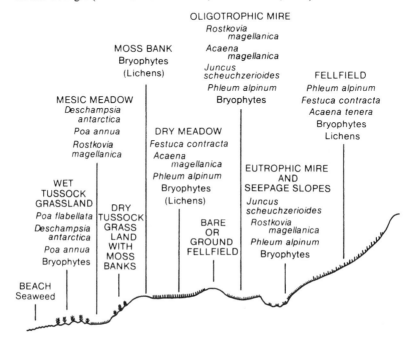

areas such as raised beaches and ridgetops, *P. flabellata* has been eradicated and replaced either by mossbanks or the introduced grass *Poa annua*. Swards of *D. antarctica* have been eradicated from mesic meadows in some areas and replaced by *R. magellanica* or *P. annua*. Closed swards of *A. magellanica* no longer occur in reindeer areas, and inflorescences are absent from, and leaf size considerably reduced in, *A. magellanica* growing in other communities. Finally, macrolichens (notably species of *Cladonia*, *Pseudocyphellaria* and *Usnea*) have almost been eradicated in reindeer areas.

Detailed botanical data on the relative abundance and biomass of different communities or plant species are lacking even for ungrazed areas of South Georgia. However, estimates available for peak standing crop (biomass) and production (Table 5.1) show a lack of homogeneity between tussock grassland and other communities. Hence, the area covered by tussock grassland had to be determined in order to provide an overall estimate of the relative abundance or biomass of each plant species. False colour aerial photographs of reindeer areas were not available to allow accurate estimates of tussock grassland extent. Nevertheless other data provide an initial assessment of the relative abundance of different communities and plant species on the Barff Peninsula (Figure 5.2).

> 80 aluminium snow poles were sited at regular intervals along transects in three localities on the Barff Peninsula. Ten 50×50 cm quadrats were thrown within a 5 m radius of each pole to determine community type and % cover of each plant species in the surrounding vegetation during mid-summer. Snow poles were considered to have been sited randomly in all communities other than tussock grassland. Colour photographs, from which areas of tussock grassland were mapped, were taken from mountain tops and verified that the 15 (19%) poles sited in this community type gave an approximate estimate of its representation in the total area of vegetation. As this estimate may later require modification, the % cover of each plant species was calculated in two ways: (1) using data from all 800 quadrats (Figure 5.2); (2) excluding the 150 quadrats read from, and dominant species occurring in, tussock grassland (shown later in Tables 5.2, 5.4).

Few of the 26 native vascular species achieve dominance on the Barff Peninsula (Figure 5.2). Indeed, the seven pteridophytes and one species of rush were not recorded at all, and all six forbs and the only species of sedge were very uncommon. The dwarf shrub, *Acaena tenera* is smaller and more scarce than *A. magellanica*. Most introduced species, except *Poa annua*, are restricted to the vicinity of the whaling stations (Walton,

1975). The native grasses, *Festuca contracta* and *Phleum alpinum*, and the rushes, *R. magellanica* and *Juncus scheuchzerioides*, do not appear to have been affected by reindeer grazing (Kightley & Lewis Smith, 1976) and remain widespread in several communities (Figure 5.2). Of the many species of bryophyte, only *Polytrichum* spp. and *Tortula robusta* were recorded individually in quadrats. As a group bryophytes are widespread throughout the vegetation, and their distribution has been increased to some extent by the overgrazing of vascular species.

The chemical composition of South Georgian plants has been measured in areas occupied by, and free of, reindeer (Lewis Smith & Walton, 1973; Walton & Lewis Smith, 1980; Pratt & Lewis Smith, 1982; Gunn & Walton, 1985). The seasonal nutrient dynamics of plants on South Georgia is similar to that in Arctic regions (e.g. Whitten & Cameron, 1980), but there are no estimates of digestibility of plants on South Georgia. Species rich in macronutrients include *D. antarctica*, *A.*

Figure 5.2. Estimates of the percentage cover of plant communities and species, and the frequency of occurrence of species in different communities on the Barff Peninsula, based on scores from 800 50 × 50 cm quadrats.

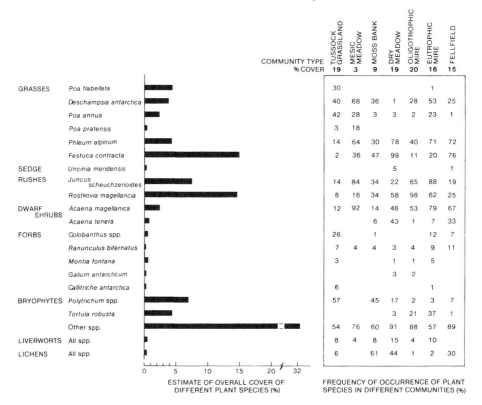

COMMUNITY TYPE	TUSSOCK GRASSLAND	MESIC MEADOW	MOSS BANK	DRY MEADOW	OLIGOTROPHIC MIRE	EUTROPHIC MIRE	FELLFIELD
% COVER	19	3	9	19	20	16	15
GRASSES *Poa flabellata*	30					1	
Deschampsia antarctica	40	68	36	1	28	53	25
Poa annua	42	28	3	3	2	23	1
Poa pratensis	3	18					
Phleum alpinum	14	64	30	78	40	71	72
Festuca contracta	2	36	47	99	11	20	76
SEDGE *Uncinia meridensis*				5			1
RUSHES *Juncus scheuchzerioides*	14	84	34	22	65	88	19
Rostkovia magellancia	8	16	34	58	98	62	25
DWARF SHRUBS *Acaena magellanica*	12	92	14	46	53	79	67
Acaena tenera			6	43	1	7	33
FORBS *Colobanthus* spp.	26		1			12	7
Ranunculus biternatus	7	4	4	3	4	9	11
Montia fontana	3			1	1	5	
Galium antarcticum					3	2	
Callitriche antarctica	6					1	
BRYOPHYTES *Polytrichum* spp.	57		45	17	2	3	7
Tortula robusta				3	21	37	1
Other spp.	54	76	60	91	88	57	89
LIVERWORTS All spp.	8	4	8	15	4	10	
LICHENS All spp.	6		61	44	1	2	30

ESTIMATE OF OVERALL COVER OF DIFFERENT PLANT SPECIES (%)

FREQUENCY OF OCCURRENCE OF PLANT SPECIES IN DIFFERENT COMMUNITIES (%)

magellanica, *P. annua* (nitrogen and phosphorus) and *P. flabellata* (soluble carbohydrate). Species of low quality include *F. contracta*, *R. magellanica* and some common bryophytes. *P. flabellata* flowers early, during September and October, whilst other species produce flowers and fruits at different times throughout the summer growing season, which ends in April or May for most species. Snow has a major effect on the growth and phenology of plants (Walton, 1982), as it does on the ecology of reindeer.

5.3   Effect of snow upon use of plant communities

In the Northern Hemisphere reindeer and caribou cannot dig through snow of more than 50–70 cm in depth (Pruitt, 1959; Formosov, 1964; Henshaw, 1968b; Lent & Knutson, 1971; La Perriere & Lent, 1977; Thing, 1977, 1981; Skogland, 1978) or through hard snow. Snow hardness has been measured by different methods, some of which have involved digging through the snow profile (e.g. Pruitt, 1959). To define plant communities that remain available to reindeer throughout South Georgian winters, I measured snow depth and hardness at minimally disturbed sites and derived exclusion values from studies using the same technique (Lent & Knutson, 1971; La Perriere & Lent, 1977; Skogland, 1978). As reindeer and caribou excavate the softest and shallowest snow profile to locate forage (La Perriere & Lent, 1977; Thing, 1977; Skogland, 1978), feeding activity was observed directly to define which communities were utilised.

> Snow depths were read at 34 poles in one locality on the Barff Peninsula during 1974 and at 80 poles in three localities during 1975, whilst snow hardness was measured using a rammsonde penetrometer (Anon., 1970) during 1975 (Leader-Williams, Scott & Pratt, 1981). To define which communities were unavailable to reindeer during 1975, data from different poles sited in the same community type were combined regardless of their aspect or height above sea level: depths of 60 cm and integrated ram hardnesses of 22 kg were assumed to exclude reindeer (cf. Skogland, 1978). Snow depths in the 1974 and 1975 winters were compared using an index of the numbers of the same 34 poles at which depths were ⩾60 cm. The use of communities at different seasons was defined by direct observation of reindeer shot whilst feeding. Selectivity of communities at different seasons was defined by comparing availability (A) and utilisation (U), using Ivlev's (1961) index of selectivity = $(U - A)/(U + A)$.

The first snow to lie on vegetated ground falls in late May or early June (Figure 5.3). During June 1975, mean snow depth and hardness readings did not exceed the theoretical reindeer exclusion values of 60 cm or 22 kg on any community (Figure 5.4). However, in July either mean snow depth or hardness, or both, exceeded the exclusion values for all communities except tussock grassland. Mean snow hardness values were generally at their highest during July, but snow depths were generally at their greatest in August. By September, mossbank and dry meadow communities were also available to reindeer. By late October 1975, only poles situated in drifts had any snow left around them. In 1974 more snow fell in August (Figure 5.3). As a result, the spring melt occurred later and vegetation was covered for longer (6.3 months) than in 1975 (5 months). Although less than the 8 months of snow experienced by barren-ground caribou (Pruitt, 1959) or the 6.3–7.6 months by Norwegian mountain reindeer (Skogland, 1978), snow lies for longer on South Georgia than might be expected from its low latitude.

Snow depth and hardness readings show that limited areas of the

Figure 5.3. Comparison of snow depths in Sörling Valley (Figure 3.13) and diet diversity (Table 5.3) in 1974 and 1975. Black and white arrows show the first and last dates on which snow lay on vegetated ground. Unbroken lines show dates of depth readings. No depth recordings were made in August 1974, but data shown with broken lines are derived from observations at Grytviken (Figure 3.13) (from Leader-Williams *et al.*, 1981).

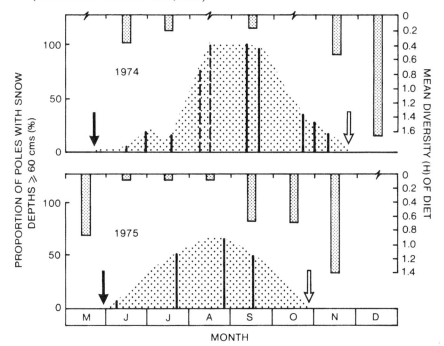

tussock grassland community remain available throughout winter months on South Georgia (Figure 5.4). Even in months when other communities are available, tussock grassland is still energetically economical (Thing, 1977) for reindeer to utilise because it is covered by relatively soft and shallow snow. Furthermore, *P. flabellata* was exposed around 50% or more of poles in that community throughout the 1975 winter. Even in the severe winter of 1974, small areas of tussock grass were visible on peninsula extremities. In contrast, vegetation was never exposed in other communities during July and August 1975, and occurred around only *c.* 10% of poles in June and September. Unsurprisingly, signs of feeding activity (Figure 5.5A) and community use (Figure 5.6) are restricted entirely to tussock grassland during winter months. Reindeer do not dig their typical circumscribed feeding craters (e.g. Pruitt, 1959; Formosov, 1964) on South Georgia. Instead they scrape snow away from the leaf bases of tussocks to expose them further (Figure 5.5B), in a slow sweeping movement made only in soft snow in the Northern Hemisphere (Thing, 1977).

Figure 5.4. Snow depth and hardness (mean ± SE) on different community types in Sörling Valley, Ocean Harbour and Hound Bay (Figure 3.13) throughout 1975 (from Leader-Williams *et al.*, 1981).

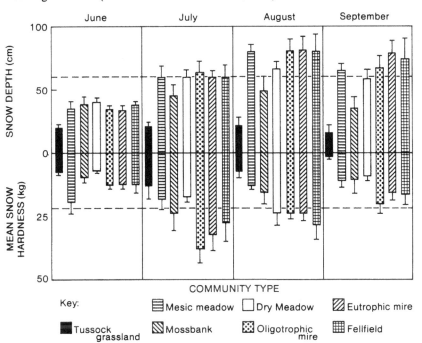

Figure 5.5. During winter reindeer forage almost exclusively in tussock grassland. In coastal areas (A) tussocks can remain exposed for much of the winter. Where snow is deeper (B) reindeer scrape around tussocks to expose them further, and they sometimes venture to steeper hillsides to forage (see Figure 7.10).

Snow depth and hardness data were read in 1975 at each of the three localities on consecutive days, and were considered homogenous between localities in each month as no major snow-falls (>15 cm) occurred between readings. There were no consistent differences between mean depth and mean snow hardness on the various community types between west (Sörling Valley) and east (Ocean Harbour, Hound Bay) localities between or within months. Data for all localities were thus combined with respect to community type and month. Analysis of variance shows that there are significant differences in snow depth and hardness between all community types: June, $F_{6,77} = 4.32$ and 2.34, $P < 0.01$ and 0.05; July, $F_{6,77} = 6.39$ and 3.87, $P < 0.001$ and 0.01; August, $F_{6,77} = 6.84$ and 4.25, $P < 0.001$ and 0.01; September, $F_{6,77} = 5.10$ and 4.00, $P < 0.001$ and 0.01. However, there are no significant differences between snow depth and hardness on all communities other than tussock grassland. Thus snow is shallower and softer on this community during each month of winter.

As snow recedes in spring, mossbank and dry meadow communities become available (Figure 5.4) and signs of feeding activity and utilisation

Figure 5.6. Use of communities at different seasons (Table 5.3) based on single observations of reindeer shot whilst feeding.

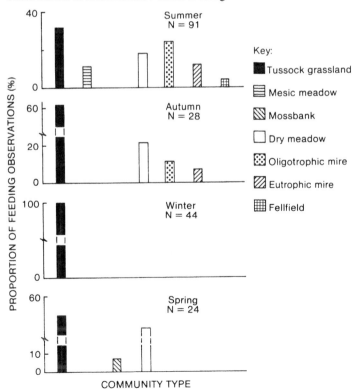

Figure 5.7. During spring and summer months reindeer select preferred forage. Fragments of lichens are sought amongst the receding snow in dry meadows during October (A), and leaves of *Acaena magellanica* are eaten from December to April (B); in both cases *Festuca contracta* is left virtually ungrazed.

are seen on them as well as on tussock grassland (Figures 5.6, 5.7A). Reindeer on South Georgia therefore follow a gradient of vegetation release from snow cover, as in the mountains of Norway (Skogland, 1980). During summer and autumn when the ground is free of snow more communities are used. In summer mesic meadow, tussock grassland and oligotrophic mire communities are selected preferentially, whereas moss-banks and fellfield are the least favoured communities (Table 5.2). In the Northern Hemisphere biting flies play a major role in determining which areas are used by reindeer and caribou in summer (e.g. White *et al.*, 1975; Skogland, 1980; 1984). Because biting flies are absent on South Georgia, it is possible to relate commmunity use directly to the availability of plants within them.

5.4   Seasonal changes in diet

Direct observations of feeding animals and examination of rumen contents are commonly used to assess amounts of different forage species eaten. However, neither method is ideal and both have inherent biases. Furthermore, the amount of lichens eaten by *Rangifer* is difficult to estimate correctly, due to their fragility and quick rate of breakdown in the rumen compared with vascular species (Courtright, 1959; Bergerud & Russell, 1964; Dearden, Hansen & Pegau, 1975; Gaare, Sørensen & White, 1977). Despite these difficulties, I present results from the examination of a small sample of direct observations and a large number of rumen samples (Leader-Williams *et al.*, 1981). First I consider the seasonal and sexual differences in the forage species eaten, and then discuss possible factors that might determine which species are selected.

A note was made of any plant species found in the mouths of shot reindeer and a sample of rumen contents was removed from each animal. Plant fragments >2 mm in size from rumen sub-samples were identified over two grids each with 100 intersection points (Leader-Williams *et al.*, 1981). The two species each of *Juncus, Acaena* and *Colobanthus*, and the introduced *Poa annua* and *P. pratensis*, were distinguished only as genera. All other monocotyledons and dicotyledons were identified at species level. In the case of *Juncus* and introduced *Poa* spp., one species is rare and most material in the samples was probably *J. scheuchzerioides* and *P. annua*. Byrophytes, lichens and seaweeds were considered only as groups.

The proportion (*p*) of each plant species in individual samples was derived by averaging the number of intersections touched on the two grids, and individual *p* values were transformed to arcsin *p* to normalise data (Snedecor & Cochran, 1967). The diversity (*H*) of individual

Table 5.2. Availability, utilisation and selectivity of plant communities on South Georgia during summer months

Community type	All community types			Without tussock grassland		
	Availability (%)	Utilisation (%)	Selectivity[a]	Availability (%)	Utilisation (%)	Selectivity[a]
Tussock grassland	19	32	+0.3			
Mesic meadow	3	11	+0.6	4	16	+0.6
Mossbank	9	0	−1.0	11	0	−1.0
Dry meadow	19	18	0	23	26	+0.1
Oligotrophic mire	20	24	+0.1	25	35	+0.2
Eutrophic mire	16	12	−0.1	19	17	−0.1
Fellfield	15	3	−0.7	18	4	−0.6

[a] Positive values indicate preferential selection, and negative values avoidance.

samples was calculated using the Shannon-Wiener diversity index (Poole, 1964), with bryophytes, lichens, seaweed, *Acaena* spp., *Colobanthus* spp., *Juncus* spp. and alien *Poa* spp. each considered as one species. Data from individual rumen samples were grouped to compare differences between: (1) different age classes; (2) the sexes; (3) adjacent months; (4) same months of different years. Single classification analysis of variance (Sokal & Rohlf, 1969) was used to compare differences in *H*, whilst differences in transformed *p* were compared with Student's *t* test. The frequency of occurrence (*f*) of species in rumens was compared by χ^2 only when *p* values were consistently <10%. An early error in mis-identification of *F. contracta* resulted in an overestimation of *p* for *D. antarctica* by <3% and an underestimation of *H* by <0.10 in samples for April to October 1975. The selectivity of forage species was assessed by comparing their availability (Figure 5.2) and utilisation (from mean proportions present in rumen samples) using Ivlev's (1961) index of selectivity (Section 5.3).

5.41 Diversity and dietary season

The diet of reindeer on South Georgia comprises relatively few plant species or species groups. In any one month, there was a maximum of only 13 species and a minimum of four, including those found only in trace quantities. In comparison, Norwegian mountain reindeer have up to 18 vascular species, as well as mosses and lichens in their rumens during summer months (Skogland, 1980).

The dietary seasons on South Georgia can be defined using differences in diversity of rumen samples (Table 5.3A). Between December 1974 and November 1975 four dietary seasons (Figure 5.8) are evident: summer – December to March; autumn – April and May; winter – June to August; spring – September and October, with a return to summer in November. The first snows fell on similar dates in 1974 and 1975 (Figure 5.3) and there was little difference in diet diversity in early winter between years (Table 5.3B). However, rumen samples were much less diverse in September and November of 1974 than of 1975. This reflected the additional month for which snow lay in 1974 (Figure 5.3), when the dietary seasons included September as a winter, and November as a spring, month. This definition of the seasons is used elsewhere in this study (e.g. Figures 4.1, 5.6, 5.10).

5.42 Species composition of the diet

There are marked differences in diversity and species composition of the diet over the annual cycle. In contrast, there are only slight

Table 5.3. *Comparisons of diversity of rumen samples (A) between months and dietary season and (B) between years*

A. Between months and dietary season

Year	Month	Mean diversity ($\bar{H}$)	Sample size	F	df	P	Dietary season
1974	June	0.39	11				Winter
				1.42	1,23	NS	
	July	0.22	14				Winter
				≈0	—	NS	
	September	0.19	19				Winter
				14.50	1,37	<0.001	
	November	0.54	20				Spring
				67.07	1,30	<0.001	
	December	1.66	12				Summer
				≈0	—	NS	
1975	January	1.67	39				Summer
				≈0	—	NS	
	February	1.67	31				Summer
				0.45	1,66	NS	
	March	1.51	37				Summer
				21.51	1,56	<0.001	
	April	1.01	21				Autumn
				0.61	1,31	NS	
	May	0.88	12				Autumn
				50.51	1,24	<0.001	
	June	0.09	14				Winter
				≈0	—	NS	
	July	0.10	9				Winter
				≈0	—	NS	
	August	0.09	13				Winter
				67.14	1,22	<0.001	
	September	0.67	11				Spring
				≈0	—	NS	
	October	0.69	13				Spring
				21.54	1,27	<0.001	
	November	1.41	16				Summer

B. Between years

Month	Mean diversity ($\bar{H}$) 1974	1975	F	df	P
June	0.39	0.09	4.82	1,23	<0.05
July	0.22	0.10	2.33	1,21	NS
September	0.19	0.67	38.75	1,27	<0.001
November	0.54	1.41	44.57	1,34	<0.001

differences between rumen samples from different age classes (calves as compared with adults and yearlings) or between those from male and female reindeer. Furthermore, diversity, rather than species composition, is the main source of difference. Thus results from the analysis of rumen samples (Figure 5.9) and from contents of reindeer mouthfuls

Figure 5.8. Species composition and diversity ($\bar{H}$) of rumen samples in different seasons in 1975. January = summer; April = autumn; August = winter; September = spring. Solid bars represent p values (%), open bars represent f values (%) (from Leader-Williams *et al.*, 1981).

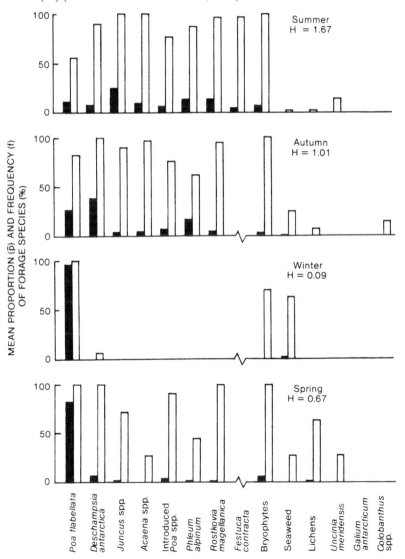

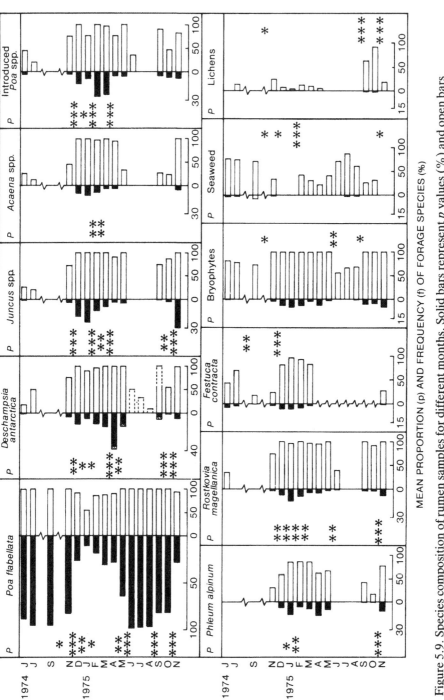

MEAN PROPORTION (p) AND FREQUENCY (f) OF FORAGE SPECIES (%)

Figure 5.9. Species composition of rumen samples for different months. Solid bars represent f values (%) and open bars represent p values (%). Significant differences between values for adjacent months are shown with $* = P < 0.05$, $** = P < 0.01$, and $*** = P < 0.001$. Broken lines and cross-hatching for *Deschampsia antarctica* represent probably small amounts of mis-identified *Festuca contracta* (from Leader-Williams et al., 1981).

(Figure 5.10) are combined with respect to age and sex, and examined in more detail for seasonal changes in species composition of the diet.

Rumen samples for single months from reindeer of different age classes or sexes showed few differences in composition or diversity (mostly $P > 0.10$). It is only when results from different seasons (Table 5.3) are examined that any differences become most apparent. Calves of 1 to 4 months of age have more diverse rumen samples ($F_{1,117} = 9.23$, $P < 0.01$) than adults and yearlings in summer months, but there are no differences in species composition (p) during these months. After the summer, there are no differences in H or p for different plant species in autumn, winter or spring months between calves of 5 to 12 months of age and adults and yearlings.

The diversity of male samples does not differ from that of females either in early summer (December, January), or in autumn and spring months. However, in both February and March (approaching and around the time of the rut) male samples are less diverse ($F_{1,25} = 4.29$ and $F_{1,31} = 4.54$, both $P < 0.05$) than those of females, and p is higher for *Poa flabellata* and lower for *Deschampsia antarctica* ($t = 2.35$ and 2.21, $df = 56$, both $P < 0.05$). Furthermore, in winter months diversity is lower ($F_{1,49} = 4.14$, $P < 0.05$) and p for *P. flabellata* is higher ($t = 2.40$, $df = 48$, $P < 0.05$) in male than in female samples.

Poa flabellata predominates in the winter diet of reindeer on South Georgia (Figure 5.9). Seaweeds washed up on beaches are an additional source of winter forage not limited by snow cover. Even though lichens may be underrepresented in rumen samples, it can be concluded that reindeer on South Georgia only seek them during spring months (Figures 5.7A, 5.9). Unlike the Northern Hemisphere, lichens on South Georgia occur in communities unavailable to reindeer in winter (Figure 5.2). Even if lichens had not been overutilised (Lindsay, 1973), they are unlikely to have been of importance in the winter diet since their typical habitats do not include tussock grassland. However, reindeer still maintain a predilection for lichens when they are available in spring months. These findings confirm earlier suggestions that reindeer on South Georgia are unusual in depending upon a single species of grass for their winter forage (Olstad, 1930; Bonner, 1958).

When snow cover ceases to limit the availability of forage species, *P. flabellata* becomes less common in rumen samples. Only six other species are found commonly in summer and autumn months (Figure 5.9). *Juncus* spp. are most abundant in early summer, introduced *Poa* spp. in late summer, and *Deschampsia antarctica* is most abundant in autumn. *Acaena* spp. (most of which occurred as lignified stem material), *Rostko-*

Figure 5.10. Species composition of reindeer mouthfuls at different seasons. A single species per mouthful is scored as 1, and three species per mouthful is scored as 0.33.

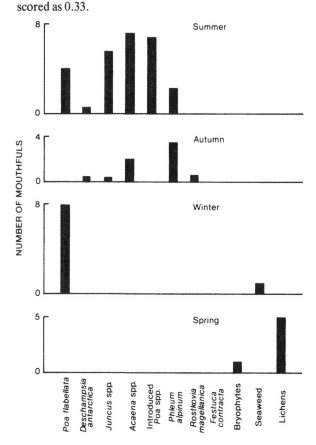

via magellanica and *Phleum alpinum* show less distinct peaks. Of the remaining species, *Colobanthus* spp., *Galium antarcticum* and *Uncinia meridensis* (none of which are shown in Figure 5.9) are found occasionally in summer months. *Festuca contracta* and bryophytes are present in small quantities throughout the year, but occur only regularly in snow-free months.

5.43   *Forage selection during summer*

During snow-free months in the Northern Hemisphere, reindeer and caribou that are not harassed by flies select forage from communities with a high standing crop of nutritious species (White *et al.*, 1975; Skogland, 1980, 1984; White & Trudell, 1980). Reindeer on South Georgia prefer to feed upon mesic meadow and tussock grassland communities during snow-free months (Table 5.2) and dominant species

Table 5.4. *Availability, utilisation and selectivity during summer months*

	All major forage species			Without species dominant in tussock grassland		
	Availability %±SE	Utilisation %±SE	Selectivity[a]	Availability %±SE	Utilisation %±SE	Selectivity[a]
Poa flabellata	4±0.2	23±2.2	+0.7			
Introduced *Poa* spp.	2±0.4	14±1.4	+0.7			
Deschampsia antarctica	4±0.7	17±1.5	+0.6			
Phleum alpinum	4±0.8	10±1.3	+0.4	5±1.0	21±2.7	+0.6
Acaena spp.	3±0.2	5±0.4	+0.3	4±0.3	14±0.9	+0.5
Juncus spp.	7±0.6	14±1.0	+0.3	9±0.7	30±2.2	+0.5
Rostkovia magellanica	14±0.9	7±0.5	−0.3	18±1.1	15±1.1	−0.1
Festuca contracta	15±1.0	4±0.4	−0.6	19±1.1	8±0.9	−0.4
Bryophytes	44±1.3	5±0.4	−0.8	43±1.5	11±0.9	−0.6

[a] Positive values indicate preferential selection, and negative values avoidance.

in these communities (Figure 5.2) are important in the diet (Figure 5.9). Comparison of the availability and utilisation of each species during snow-free months shows that *Poa flabellata*, introduced *Poa* spp. and *Deschampsia antarctica* are the most preferred species, whilst *Rostkovia magellanica*, *Festuca contracta* and bryophytes are generally avoided (Table 5.4). Even if species dominant in tussock grassland are excluded from the analysis in case of error in estimating the percentage cover of tussock grassland, the ranking order of selection or avoidance of the remaining forage species does not alter.

Measures of macronutrient contents, structural components and calorific value are available for most major forage species throughout the growing season. In addition, measures of soluble carbohydrate content are available for *Poa flabellata* (Gunn & Walton, 1985). As the diet of reindeer on South Georgia is dominated by grasses, secondary chemical components are probably of no importance (cf. Kuropat & Bryant, 1983). It appeared that nitrogen (Figure 5.11), and to a lesser extent, phosphorus concentrations are an important determinant of forage species occurring commonly in rumen samples (Leader-Williams *et al.*, 1981). Correlations of the index of selectivity of each species against chemical composition confirmed the importance of nitrogen (Figure 5.12). There are no other significant positive or negative correlations between other macronutrients, structural components or calorific value of forage species and their importance in the diet (Table 5.5). As in the Northern Hemisphere (e.g. Klein, 1970b; Skogland, 1980, 1984), reindeer on South Georgia generally select the most nutritious forage available, especially for its nitrogen (and therefore crude protein) content.

Poa flabellata stands out as the one species which is selected preferentially yet is of low nitrogen content (Figure 5.12). In the Northern Hemisphere, *Rangifer* selects nutritious forage species with a high overall standing crop to maximise forage intake during a given feeding period (White *et al.*, 1975; Skogland, 1980, 1984; Trudell & White, 1981). Estimates of standing crop and production are only available within ungrazed communities on South Georgia (Table 5.1). Even though *P. flabellata* is now locally overgrazed, it still has the highest overall standing crop of any community in the reindeer areas. Grazing mammals also select forage species for their high carbohydrate (and therefore energy) content (Arnold, 1964; White, 1983), and *P. flabellata* has exceptionally high concentrations of fructans in its shoots (Gunn & Walton, 1985). Unfortunately no measures of carbohydrate content are available for

Figure 5.11. Comparison between mean proportions of forage species present in rumen samples and their macronutrient content during different snow-free months. Values of % N are shown for *Poa annua*, *Acaena magellanica*, *Juncus scheuchzerioides* and *Tortula robusta* (redrawn from Leader-Williams *et al.*, 1981).

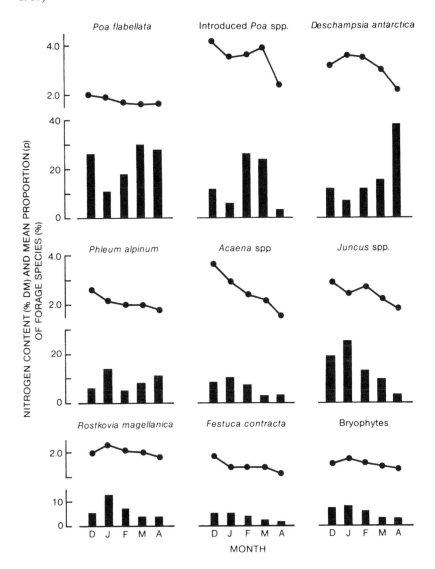

Table 5.5. *Results of correlations between the index of selectivity and chemical composition of forage species, as shown for* N *in Figure 5.12*

Chemical component	Spearman rank correlation coefficient	N	P
N	0.63	9	<0.05
P	0.58	9	NS
Ca	−0.45	8	NS
Ash	0.34	8	NS
Holocellulose	−0.25	8	NS
Crude fibre	0.02	8	NS
Calorific value	0.26	8	NS

other South Georgian species. In the absence of these data, it is not yet possible to determine whether the high standing crop or the high carbohydrate content, or both, determine the high selection reindeer show for *P. flabellata* during snow-free months. Both factors may be important, especially near the end of the growing season of other major forage species. In autumn, reindeer show both a marked switch in the use of *P. flabellata*, and a decrease in diet diversity well in advance of the limitation of vegetation communities by snow cover (Figure 5.4, 5.9). Interestingly, snow rarely limits forage on the Falkland Islands and sheep also grazed *P. flabellata* (before they eliminated it) more in autumn than

Figure 5.12. Correlation between the index of selectivity and mean chemical composition of forage species during snow-free months. Correlation includes data from species dominant in tussock grassland communities (Table 5.4), and uses the mean value for % N measured monthly from December to April (Figure 5.11).

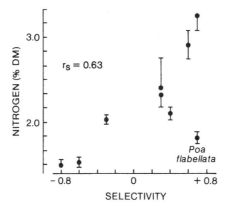

at other times of year (Davies, 1939). Hence *P. flabellata* is an important forage even when other species are available, and it would be worthwhile to study further the factors which determine its selection.

5.5   Nutritional limitations

In the Northern Hemisphere, reindeer and caribou face a 'boom-and-bust' environment, and have adapted to it in many ways. Captive reindeer and caribou voluntarily reduce their intake of food and of energy and nitrogen in winter (McEwan, 1968; McEwan & Whitehead, 1970). Furthermore, free-living reindeer which depend largely upon lichens (a forage poor in nitrogen, but rich in energy) can sustain a negative nitrogen balance for that period (Syrälä-Qvist & Salonen, 1983). On South Georgia, also, a forage species poor in nitrogen predominates in the winter diet of reindeer. The seasonal availability of nitrogen in forage eaten by reindeer will now be examined as an indication of nutritional limitations acting on South Georgia. I present data on changes in weight and in nitrogen concentrations of rumen contents.

The rumen and reticulum was weighed full and empty to obtain weight of rumen contents by subtraction. One sub-sample was oven-dried to obtain % dry matter (DM) and thus dry weight of rumen contents. A second sub-sample was sieved and fragments >2 mm were milled, dried and analysed for nitrogen concentrations (% DM) (Allen, Grimshaw, Parkinson & Quarmby, 1974) as an index of the quality of forage consumed (Klein, 1962; Staines, Crisp & Parrish, 1982). Nitrogen concentrations in rumens of different age groups and of each sex were compared with the Kruskal-Wallis one-way analysis of variance (ANOVA) (Siegel, 1956). Multiplication of the % N×dry weight of rumen contents gave an estimate of the total weight of ruminal N derived from the plants eaten. Sine waves (Leader-Williams & Ricketts, 1982a) were used to compare inter-relationships of measures of diversity, weight of rumen contents and nitrogen concentrations to the age and month of collection of each animal. As diet diversity and % N fluctuated by season but did not show any increase with age, a simple sine wave formula was fitted to these data. The weight of rumen contents and total ruminal N increased with age and showed seasonal fluctuations. Thus a sine wave formula incorporating a growth curve was fitted to these data. The sine wave formula allowed estimation of the asymptotic value (A), proportional fluctuation about the curve (C), and phase adjustment of the sine wave (Φ). Differences in A, C and Φ were tested by fitting common parameters and performing an analysis of

Table 5.6. *Seasonal changes in diversity, nitrogen (% DM), weight rumen contents (kg) and total ruminal N (g DM). Maximum values occur at Φ+3 months and minimum values at Φ+9 months relative to mid-November. Braces indicate common values*

Parameter and sex	Growth rate (k)	Asymptotic value (A)	Phase adjustment (Φ)	Proportional fluctuation (C)	No. of weighted means
Dietary diversity					
Male	—	0.82	{−1.0}	{98}	50
Female		0.91			79
Nitrogen content of rumen					
Male	—	{2.5}	{−1.0}	{11}	49
Female					78
Rumen weight					
Male	{0.068}	17.0	{1.0}	{12}	49
Female		13.5			78
Total ruminal nitrogen					
Male	{0.084}	{39}	{−1.0}	{22}	43
Female					65

variance. More details of the sine wave formula and rationale of its use
are given in Chapter 6 and in Leader-Williams & Ricketts (1982a).

Female red deer select more nutritious forage than males, even though
plant species found in their rumens differed little (Staines *et al.*, 1982). In
reindeer, there are slight differences in species composition and diversity
of rumen samples (Section 5.42), but nitrogen concentrations in the
rumen do not differ between males and females, or between calves,
yearlings and adults. Therefore, female reindeer and calves do not
appear to select more nutritious forage than males or older reindeer on
South Georgia. As with species composition, differences in nitrogen
concentrations arise mainly from seasonal changes (Figure 5.13). Sine
waves for changes in diversity and nitrogen concentrations confirm these
conclusions over the annual cycle (Table 5.6). Males have slightly less
diverse diets than females, but nitrogen concentrations do not differ.
Both parameters fluctuate in phase with each other, and by the same
proportions in each sex. The average nitrogen concentration of forage
consumed by reindeer is 2.5% (or 15.6% crude protein). Maximum
values of 2.8% (17.3% crude protein) are reached in January and
minimum values of 2.2% (13.9% crude protein) are reached in July.

Weights of rumen contents are heavier in males than in females, but
fluctuate in phase and by the same proportion in both sexes (Table 5.6;
Figure 5.14). The rumen is at its heaviest two months later than the
maxima of nitrogen concentrations and total ruminal nitrogen, suggesting that diet quality begins to reduce before intake becomes restricted. In
spite of sex differences between the weights of rumen contents, sine
waves show that there is no difference in the total ruminal nitrogen

Figure 5.13. Changes in the nitrogen and crude protein (N × 6.25)
concentrations in rumen contents of reindeer at different seasons showing basic
data of mean ± SE for each month with sample sizes, and the sine wave fitted to
these data (Table 5.6).

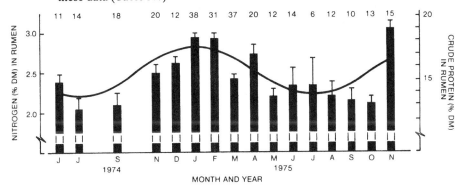

Figure 5.14. Phasing and proportional fluctuations in weight of the rumen and its contents, and of the total ruminal nitrogen over one annual cycle.

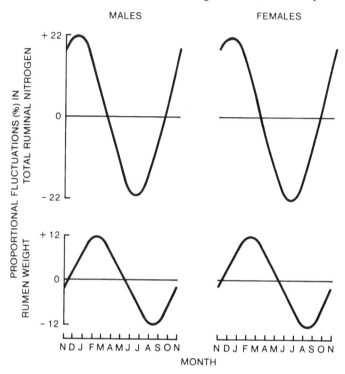

available to males and females. In this respect reindeer are similar to red deer (Staines *et al.*, 1982).

In conclusion, the quantity and quality of forage eaten at different seasons differs markedly because only *Poa flabellata* is available during winter months, and because reindeer voluntarily reduce their food intake at that period. However, the quality of food eaten and the availability of forage-derived nitrogen in the rumen shows only slight differences between male and female reindeer. This result is surprising and contradictory to predictions for dimorphic cervids (Clutton-Brock *et al.*, 1982). No data appear to be available from other studies of *Rangifer*'s feeding ecology to determine whether my result is a consequence either of the unusual situation in which reindeer find themselves on South Georgia, or of the tendency of the species to forage in large mixed groups (Figure 4.1).

Kruskall-Wallis ANOVA shows that there is no difference in % N of rumen contents between adult, yearling or calf males and females in any month or season. Thus monthly changes in % N are shown for all reindeer in Figure 5.13. Sine waves provide an adequate description of change in parameters affected by season because they significantly ($P <$ 0.001) reduce the residual sum of squares compared with that derived from fitting a growth curve alone (Leader-Williams & Ricketts, 1982a): e.g. for total ruminal N, $F_{2,40} = 16.71$ in males and $F_{2,62} = 11.32$ in females, as illustrated for % N in Figure 5.13. Analysis of variance shows that values of A are common to males and females for % N ($F_{1,123} = 0.86$), but differ for diversity ($F_{1,125} = 4.23$, $P < 0.05$); values of C and Φ are common to both sexes for each parameter (all $P > 0.10$). Values of A for rumen contents differ between males and females ($F_{1,103} = 22.13$, $P < 0.001$) but those for total ruminal N are common to both sexes ($F_{1,102} = 1.43$, $P > 0.10$); values of C and Φ are common to both sexes for each parameter (all $P > 0.10$). Diet diversity, % N and total ruminal N have common values of Φ ($P > 0.10$) and have maximum and minimum values two months earlier than those for weight of rumen contents.

5.6   Feeding ecology at high latitudes

Reindeer and caribou have a wide distribution through a variety of different habitats at latitudes generally higher than other species of temperate cervid (Table 1.2). At lower latitudes the feeding ecology of *Rangifer* is characterised by their dependence on the lichen-based food niche (Klein, 1980a; 1986). This niche is unoccupied by all other species of northern cervid (moose: Peterson, 1955; Crête & Jordan, 1981; red deer: Jensen, 1968; Dzieciolowski, 1970; wapiti: McCullough, 1969; Houston, 1982; white-tailed deer: White, 1961; Short, 1971; black-tailed deer: Lovaas, 1958; Anderson, Snyder & Brown, 1965; roe deer: Prior, 1968; Gebczynska, 1980; fallow deer: Jackson, 1974; Chapman & Chapman, 1975), and by other northern ungulates (muskox, mountain sheep: Tener, 1965; Geist, 1971; White *et al.*, 1981) whose range overlaps with that of *Rangifer*. In adapting to this niche, reindeer and caribou have evolved more efficient digging capabilities than other cervids (Pruitt, 1959; Kelsall & Prescott, 1971) and ungulates (Lent & Knutson, 1971) to reach lichens through snow in winter.

Indeed, reindeer and their dependence upon lichens had almost become dogma prior to studies of the species at their most northerly latitudes. It is now clear that there is a latitudinal and altitudinal gradation of winter diet (Klein, 1986). This is determined to a large extent by a combination of food availability, topography and snow depths, as summarised diagrammatically in Figure 5.15. At southerly latitudes, woodland reindeer and caribou have short altitudinal migrations and depend largely upon arboreal lichens in winter (e.g. Edwards &

Ritcey, 1960; Bergerud, 1972; Bloomfield, 1980; Pulliainen, 1980). Reindeer living in alpine tundra also move along altitudinal gradients, but forage for terrestrial lichens in areas where snow is softest and shallowest (e.g. Skogland, 1975, 1978, 1984). Many barren-ground and Alaskan caribou, and many tundra reindeer populations migrate huge distances to overwinter in the taiga where they too forage for terrestrial lichens (e.g. Scotter, 1967; Kelsall, 1968; Skoog, 1968; Miller, 1976; Geller & Borzhanov, 1984; Thomas & Hervieux, 1986). However, some continental populations of tundra reindeer do actually overwinter in the tundra, and their winter diets include higher proportions of sedges and shrubs than of lichens (e.g. Michurin & Vakhtina, 1968). It is in the High Arctic, though, where insular populations of tundra and Spitzbergen reindeer and of barren-ground and Peary caribou are found, that lichens are of least importance in the diet. Here short altitudinal, and sometimes longer inter-island movements on sea-ice occur, and reindeer and caribou depend largely upon graminoids or moss in winter (e.g. Kischinskii, 1971; Hjeljord, 1975; Shank, Wilkinson & Penner, 1978; Punsvik, Syvertsen & Staaland, 1980; Miller, Edmonds & Gunn, 1982; Thomas & Edmonds, 1983; Thing, 1984). Introduced reindeer on South Georgia are, by force of circumstance, even more dependent upon grass in winter (Figure 5.16), and upon a single species at that. In spite of the predominance of *Poa flabellata* in their winter diet, concentrations of forage-derived nitrogen on South Georgia do not differ greatly from those for natural populations from which data are available in the Northern Hemisphere (Klein & Schønheyder, 1970; Reimers, Klein & Sørumgard, 1983; Thing, 1984).

Figure 5.15. Diagrammatic representation of the range of reindeer and caribou habitats in the Northern Hemisphere, together with the major winter food in each habitat, and the type of migration undertaken (the arrows indicate relative lengths of migration and whether altitudinal or longitudinal). Sources of data are given in the text.

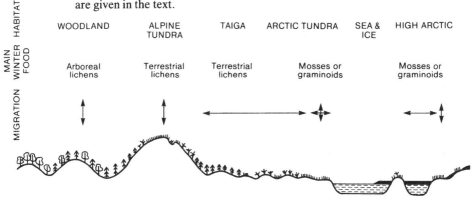

The differences in the broad patterns of summer diet are less striking than those in winter diet across the range of natural habitats which reindeer and caribou occupy (Figure 5.16). Terrestrial lichens are an important component of even the summer diet in more southerly

Figure 5.16. Latitudinal differences in caribou diets from the woodlands to the High Arctic of the Nearctic region, with comparative data from reindeer in the Southern Hemisphere. (Data from: Shank *et al.*, 1978; Thing, 1984; Skoog, 1968 and White *et al.*, 1975; Banfield, 1954 and Scotter, 1967; Bergerud, 1972; this study.)

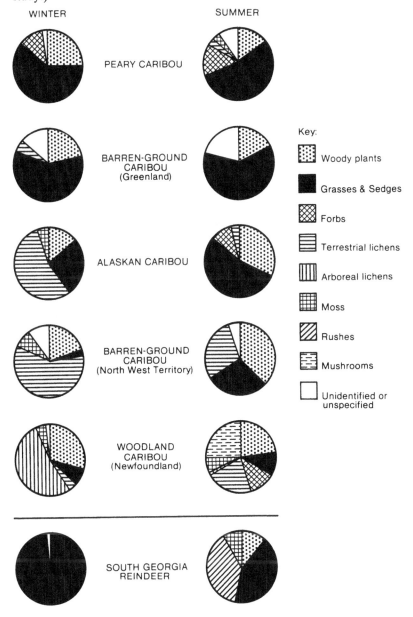

latitudes, as are also mushrooms. However, *Rangifer* can be classified as mixed or intermediate feeders (cf. Hofmann & Stewart, 1972), eating both grass and browse in summer in each habitat they occupy (Skogland, 1980, 1984b; White *et al.*, 1981; Trudell & White, 1981). In this respect reindeer in the unusual habitat of South Georgia do not differ substantially from their Northern Hemisphere counterparts (Figure 5.16).

In conclusion, reindeer on South Georgia conform to the generalisations that *Rangifer* is an adaptable mixed feeder, selecting both for quality and quantity of forage. Furthermore they demonstrate that *Rangifer* can survive without lichens in winter in extreme ecological conditions. Surprisingly, however, there are few differences in the feeding ecology of male and female reindeer.

5.7   Summary

1. There is a species-poor vascular flora on South Georgia, comprising 25 native taxa and a few introduced species. The vegetation is dominated by coastal tussock grass, *Poa flabellata*, a species of high biomass and productivity.
2. Snow cover limits the choice of forage available for reindeer to only the tussock grassland community for up to three months of winter, and other forage species remain unavailable for up to six months.
3. Reindeer on South Georgia are unusual because they depend almost exclusively upon a single species of grass during winter.
4. During summer months, reindeer also select those forage species high in nitrogen and phosphorus, notably *Deschampsia antarctica*, *Acaena magellanica* and the introduced *Poa annua*. However, other common species like rushes, though not selected, also make up a large part of the diet.
5. Concentrations of forage-derived nitrogen in the rumens of reindeer and their general patterns of feeding in summer on South Georgia do not differ from the Northern Hemisphere. The quality of forage eaten shows few differences between males and females.
6. Lichens are unimportant in the diet of reindeer on South Georgia, and demonstrate further that *Rangifer* is an adaptable feeder.

6

Growth and body condition

The females are horned as well as the males, which is proper to this order of
quadrupeds, but the horns of the females are more slender than those of the
other sex.
Carl Linnaeus (1732)

6.1 Introduction

Whilst reindeer and caribou are unique amongst cervids in that
females possess antlers, their size and shape differs between males and
females. Furthermore, there are sexual differences in other measures of
body size and weight in *Rangifer*, which are amongst the most dimorphic
of contemporary cervids (Clutton-Brock *et al.*, 1982). Studies of patterns
of growth and seasonal changes in body condition amongst *Rangifer* have
involved both captive animals and free-living populations (e.g. Krebs &
Cowan, 1962; McEwan & Wood, 1966; McEwan, 1968, 1975; Reimers,
1972, 1983c; Dauphiné, 1976; Krog *et al.*, 1976; Timisjärvi *et al.*, 1982;
Reimers *et al.*, 1983; Tyler, 1987a,b). These studies established the
existence both of seasonal patterns of growth and a voluntary reduction
in food intake in winter. They also demonstrated that reindeer and
caribou show wide variations in body size and condition, with respect to
sub-species, locality and season. Various explanations for these differences have been advanced and include differences in quality of summer
range, in levels of harassment by biting flies or by predators and in
activity patterns (Klein, 1968; Reimers, 1980, 1983c). However, it was
unclear whether males and females have similar growth rates or to which
seasonal factors the changes in weight and condition are primarily
related.

In this chapter I compare the pattern of antler (6.2) and skeletal (6.3)
growth and the timing and magnitude of annual changes in weight (6.4)
and condition (6.5) of male and female reindeer on South Georgia. I use
the data in this chapter to question whether males invest more heavily in

antler growth relative to their body weight than do females. I also ask whether or not there is a dimorphism in growth rates and in seasonal patterns of weight loss in males and females, and the extent to which these might be related to the costs of reproduction in each sex (6.6).

6.2   Antler growth

Antlers are amongst the fastest growing tissues in the animal kingdom and their structure and function have fascinated biologists since the time of Aristotle (Goss, 1983). *Rangifer* females and calves regularly grow antlers and older males have complex antlers (Figures 4.7, 6.1). This complexity frequently includes a palmated brow tine and intricate secondary branching on the main tines (Pruitt, 1966; Bubenik, 1975a; Bergerud, 1976; Miller, 1986). Furthermore, the antlers of *Rangifer* are sometimes asymmetrical in both shape and structure between left and

Figure 6.1. Male (A) and female (B) reindeer show considerable dimorphism in antler shape and in body size and weight. The antlers of adult males are complex, highly branched and frequently asymmetrical between sides. A palmated brow tine is also often present on one side with a reduced brow tine on the other (A and C). The antlers of females are less complex (B and D), whilst those of calves are usually simple and unbranched. Males attain peak weights in March (A) and lose condition rapidly during the rut when they become exhausted (C). In contrast females are at their worst condition after calving and during lactation (D).

right sides (Davis, 1973; Goss, 1980; Miller, 1986). In this section, I examine the patterns and differences in growth of antlers of each sex.

Antler state was classified into velvet, cleaning and hard categories (Chapter 4). Antler height was calculated as the mean height of both sides, measured in straight line distances inside the medial curve of the antlers, from the casting burr to the tip. Antler weight was measured as the combined weight of both sides, after antlers had been sawn off just below the casting burr. The complexity of the antlers was assessed only for cleaning or hard antlers by counting the number of main branches or tines (including a brow tine if present) on each antler, and by noting the presence of secondary branching and of dominant or reduced brow tines. Reindeer were considered to have asymmetrical antlers if left and right antlers had either a different number of main tines or a different pattern of secondary branching. As many reindeer had asymmetrical antlers, the number of main tines for each reindeer was summed for left and right sides. The frequency of occurrence of secondary branching and of brow tines was calculated from both left and right antlers of each reindeer.

Male and female calves usually grow simple antlers which consist of a single tine on each side (Figures 4.17, 6.1B, 6.2), although some (27%)

Figure 6.2. Number ± SE of main tines and frequency of occurrence of asymmetry between left and right sides, of secondary branching on main tines and of brow tines in male and female reindeer of different ages. Due to the common occurrence of asymmetry, numbers of main tines for each reindeer are summed for both sides; the occurrence of secondary branching and of brow tines are shown for each antler. Thus N for (C) and (D) = $2 \times N$ for (A) and (B).

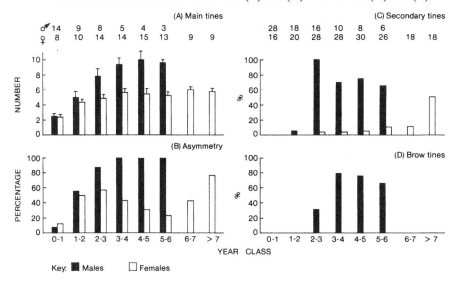

calves have a single branch on one or both sides. Yearling reindeer grow more branched antlers and show more asymmetry between sides than calves, but there is still no difference in complexity between the sexes. However, from 2 years of age male antlers become increasingly asymmetrical and more complex than those of females. Males grow more main tines and secondary branching and develop complex brow tines, which are absent in reindeer females on South Georgia (Figure 6.2). Hence there is a dimorphism in antler structure between males and females (cf. Bubenik, 1975b; Miller, 1986).

Kruskal-Wallis ANOVA, corrected for ties, shows that the number of tines increases with age in both males ($H = 31.44$, $df = 5$, $P < 0.001$) and in females ($H = 25.93$, $df = 7$, $P < 0.001$). Mann-Whitney U tests show there are no differences (both $P > 0.10$) in numbers of tines between males and females in calves ($U = 53$, $n_1 = 8$, $n_2 = 14$) or in yearling reindeer ($U = 38.5$, $n_1 = 9$, $n_2 = 10$). However, in older reindeer males grow more tines than females (2–3 years: $U = 18$, $n_1 = 8$, $n_2 = 14$; 3–4 years: $U = 5.5$, $n_1 = 5$, $n_2 = 14$, both $P < 0.02$; 4–5years: $U = 2.5$, $n_1 = 4$, $n_2 = 15$; 5–6 years: $U = 0$, $n_1 = 3$, $n_2 = 13$, both $P < 0.002$).

The height and weight of antlers also differs between the sexes (Figure 6.1). Obviously the relationship between antler weight and age in both sexes is discontinuous because of annual casting and regrowth of each antler, as shown for male reindeer only in Figure 6.3. The antlers of females show similar growth patterns but are smaller. Antler growth whilst in velvet is rapid but antlers lose weight during cleaning. This loss cannot be attributed to sampling artefact as antler lengths do not decrease during this period, but may be due to fluid loss as rapidly growing tissue becomes converted to bone. Thus antler size of each sex is compared using only hard antler weights. In calves these do not differ between the sexes. In reindeer over one year, the weight increase of hard antlers with age is not linear (Figure 6.3). Comparison of simple regression lines of $\log_e$ hard antler weight on total body weight, and multiple regression lines on total body weight and year class show that body weight is the most important factor affecting hard antler weight (Figure 6.4; cf. Huxley, 1931). Simple regression lines predict asymptotic antler weights of 2.4 kg and 0.2 kg for males and females, using the maximum total body weight for each sex. Therefore, antler weight comprises a higher proportion of total weight in male, than in female reindeer (Figure 6.5).

The dimorphism in complexity and in the weight of antlers relative to body weight in male and female reindeer raises the question of possible differences in the function of their antlers. The principal function of antlers in male *Rangifer* is for use as visual cues and in fights between competing males (Figure 4.2A; Pruitt, 1960a; Espmark, 1964a; Bergerud, 1974a; Bubenik, 1975b; Prowse, Trilling & Luick, 1980), as is the case in other cervids (Clutton-Brock, 1982). However, the functions of the smaller antlers of female reindeer and calves are less clear. They are most commonly suggested to confer an advantage in social rank when males are antlerless, and to permit females and calves access to feeding craters in winter (Espmark, 1964b; Henshaw, 1968a; Bubenik, 1975b). Comparisons between the social life of *Rangifer* and other cervids may parallel differences between genera of African antelope in which females are horned or polled (Packer, 1983). Among antelopes, females tend to have horns in species which form large mixed groups and where calves are precocious. A similar dimorphism in horn structure occurs between males and females in these genera, and horns have a less offensive function in females than in males, which may also include protection of calves from predators. Female *Rangifer* only carry hard antlers for short periods after calving, and both females and calves have them during the

Figure 6.3. Changes in antler weight in male reindeer. When two classes of antler were present in one month, mean weights for both are shown separately. Sample sizes are shown in Figures 4.16, 4.18. Curves for growth of velvet antlers and decline in weight from velvet to hard antlers are drawn freehand; horizontal lines show mean hard antler weight for each year class. Broken lines indicate where weights of velvet antlers in calves were not measured, and where data in old males were insufficient (from Leader-Williams & Ricketts, 1982a).

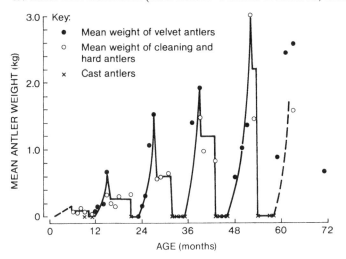

Figure 6.4. Regressions of log$_e$ hard antler weight on body weight for male and female reindeer (data from Leader-Williams & Ricketts, 1982a).

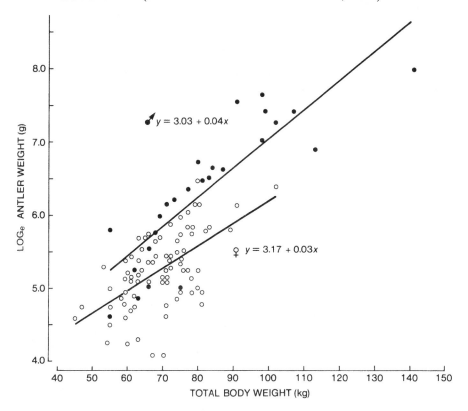

Figure 6.5. Percentage of the asymptotic total weight attained by body components in male and female reindeer (data from Leader-Williams & Ricketts, 1982a).

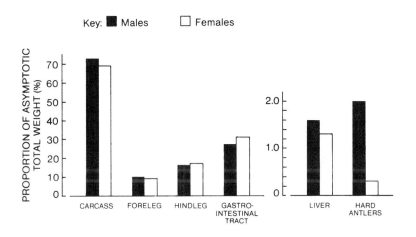

following winter, when antlers could be used offensively against predators. Clearly reindeer on South Georgia cannot be used to test this possibility, and a full explanation of the function of antlers in females and calves is still awaited from the Northern Hemisphere. However, differences in the numbers of reindeer of each sex and age class with antlers that were broken (Figure 6.6) tend to confirm that females and calves use their antlers less offensively than males, and to mirror results obtained from antelopes (Bubenik, 1975b; Packer, 1983).

Hard antler weights of reindeer calves do not differ between males and females ($t = 0.55$, $df = 17$, $P > 0.10$). The reduction in residual variance between simple regression lines of $\log_e$ antler weight on body weight, and multiple regression lines of $\log_e$ antler weight on body weight and age class is not significant ($F_{1,100} = 3.68$, $P > 0.10$). Simple regression lines of $\log_e$ antler weight on body weight for each sex are significant (males: $N = 25$, $r = 0.86$; females: $N = 79$, $r = 0.57$, both $P < 0.01$). There is no difference between the slopes of regression lines for each sex ($F_{1,100} = 2.11$, $P > 0.10$), but the intercepts do differ ($F_{1,101} = 30.52$, $P < 0.001$). Because male and female reindeer carry hard antlers when total body weights are at their maximum, asymptotic antler weights are predicted with (asymptotic total weight + [asymptotic total weight× proportional fluctuation]) (see in 6.4). There is a difference in numbers of male and female reindeer with broken antlers ($\chi^2 = 4.08$, $P < 0.05$). Several males had breaks on both left and right sides and there is a larger difference in numbers of broken antlers found in males, females and calves ($\chi^2 = 11.37$, $df = 2$, $P < 0.01$).

6.3   Skeletal growth

Skeletal measurements like mandible length and hindfoot length are longer in male *Rangifer* than in females (e.g. Bergerud, 1964a;

Figure 6.6. Number of reindeer with broken antlers, and numbers of broken antlers amongst males, females and calves. * indicates a difference of $P < 0.05$ and ** a difference of $P < 0.01$.

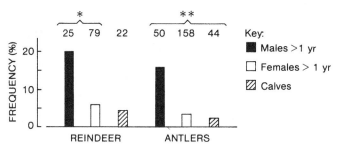

Kelsall, 1968; Reimers, 1972; Dauphiné, 1976; Krog *et al.*, 1976). However, standard growth curves that estimate parameters such as asymptotic size and growth rate have rarely been fitted to growth data from *Rangifer* (Krebs & Cowan, 1962; McEwan & Wood, 1966; McEwan, 1968, 1975). In this section, I describe patterns of skeletal growth in reindeer using an analytical technique that compares biological parameters derived from growth curves (Leader-Williams & Ricketts, 1982a).

> The growth of the skeleton was assessed by taking the following measurements from all shot reindeer, as described fully in Leader-Williams & Ricketts (1982a): (1) crown–tail length: the distance from the nuchal crest along the spine to the tail base; (2) jaw length: the straight-line distance from the point of the right mandible to the gum margin at the lingual side of the canine tooth; (3) femur length: the straight-line distance from the lateral tuberosity to the lateral malleolus after dissecting out the left femur; (4) hindfoot length: the straight line distance from the point of the hock to the horn line of the hoof. In addition, the crown–tail length of all sexed foetuses was measured, and the foetuses assigned an age from −1 to −6 months, according to the date of collection of the mother.
>
> Growth curves of the form $Y = A - B^{(-kt)}$ (Brody, 1945), where $Y =$ mean length at t months, $A =$ asymptotic length, $B =$ integration constant, and $k =$ growth rate, were fitted to length on age data by a weighted least squares technique (Leader-Williams & Ricketts, 1982a). In these analyses, k represents the decline in length per unit time as the asymptotic length is approached. The data for each parameter appeared to show growth in two phases, rapid growth up to 12 months of age and slower growth thereafter. Therefore, growth curves were fitted separately to data of 0 to 12 months, and of over 12 months, for each sex: in each case this resulted in a significant ($P < 0.001$) reduction in the residual sum of squares (tested by analysis of variance) compared with a single growth curve. As foetal data were available for crown–tail lengths, a generalised logistic growth curve (Richards, 1959) was fitted to this parameter from −6 to 12 months of age. Differences in growth rates or asymptotic lengths were assessed by fitting common parameters and performing an analysis of variance.

Crown–tail length increases logistically from early foetal life to 12 months of age (Figure 6.7), and shows the form of a classical sigmoid growth curve. The point of inflection of curves for males and females occurs at birth. At 12 months of age, males and females have similar estimated lengths and growth rates for all skeletal measurements except hindfoot length (Table 6.1). By this age, males and females have already

Table 6.1. *Skeletal growth in reindeer, as represented by crown–tail, femur, hindfoot (all in cm) and jaw (mm) lengths. Braces indicate common values (from Leader-Williams & Ricketts, 1982a)*

Measurement and sex	Growth up to 12 months of age			Growth after 12 months of age		
	Growth rate (k)	Asymptotic length (A)	No. of weighted means	Growth rate (k)	Asymptotic length (A)	No. and of weighted means
Crown–tail						
Male	{0.49}	{100}	19	{0.047}	123	38
Female			15		114	71
Jaw						
Male	{0.30}	{208}	13	{0.037}	271	38
Female			9		241	71
Femur						
Male	{0.42}	{22.5}	13	{0.083}	27.5	39
Female			9		25.5	71
Hindfoot						
Male	{0.43}	38	13	{0.057}	44	39
Female		37	9		41	71

achieved much (80–90%) of their skeletal growth (Figure 6.8). Growth rates for all skeletal components are the same in males and females over 1 year old. However, asymptotic lengths are consistently larger in males (Figure 6.7, Table 6.1). These results confirm the occurrence of a sexual dimorphism in skeletal size, similar to the growth patterns observed in *Rangifer* elsewhere (e.g. McEwan, 1968, 1975; Kelsall, 1968). Although males are larger than females, male reindeer approach their asymptotic length at the same rate as females.

Figure 6.7. Growth in mean crown–tail length of male and female reindeer. Fitted curves from −6 to 12 months of age are logistic (arrow indicates time of birth); growth in reindeer older than 12 months is described by a standard growth curve (from Leader-Williams & Ricketts, 1982a).

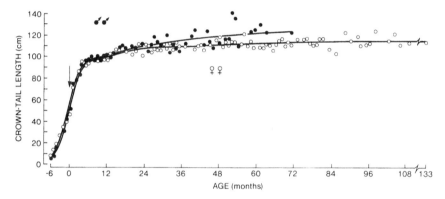

Figure 6.8. Proportion of skeletal and body growth achieved by male and female reindeer at birth and one year of age, as represented by crown–tail length and total weight.

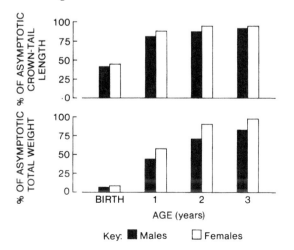

Analysis of variance shows that there is no difference in growth rate (k) ($F_{2,28} = 0.98$) or in the estimate (A) of crown–tail length ($F_{1,30} = 0.06$) at 12 months of age between the sexes (both $P > 0.10$). From birth to 12 months, values of k and A do not differ (all $P > 0.10$) between the sexes for jaw length ($F_{1,16} = 0.04$ and $F_{1,18} = 1.40$) or femur length ($F_{1,16} = 0.96$ and $F_{1,18} = 2.83$). Although values of k for hindfoot length do not differ ($F_{1,16} = 1.12$), values of A differ ($F_{1,18} = 5.10$, $P < 0.05$) at 12 months.

In reindeer over 12 months, values of k do not differ (all $P > 0.10$) between sexes for crown–tail length ($F_{1,103} = 1.10$), jaw length ($F_{1,103} = 0.15$), femur length ($F_{1,104} = 0.81$) or hindfoot length ($F_{1,104} = 0.45$). However, the adult asymptote (A) differed (all $P < 0.001$) between the sexes for all of crown–tail length ($F_{1,105} = 33.75$), jaw length ($F_{1,105} = 111.07$), femur length ($F_{1,106} = 155.98$) and hindfoot length ($F_{1,106} = 90.52$).

6.4   Body and organ growth

Body and organ weights in reindeer and caribou show patterns both of growth and of seasonal fluctuation (e.g. McEwan, 1968; Dauphiné, 1976; Krog *et al.*, 1976; Timisjärvi *et al.*, 1982; Reimers, 1983c; Reimers *et al.*, 1983; Tyler, 1987a,b). Standard growth curves have been fitted to maximum and minimum body weights in each annual cycle (McEwan & Wood, 1966; McEwan, 1968), but sine waves provide a better description of both growth and seasonal fluctuations (Moen, 1973, 1978). In this section I describe changes in body and organ weights, using methods that compare biological parameters derived from both growth curves and sine waves (Leader-Williams & Ricketts, 1982a).

Changes in body and organ weights were assessed by taking the following measurements from all shot reindeer, as described fully in Leader-Williams & Ricketts (1982a): (1) total weight: weight of unbled reindeer, excepting loss of blood from the shot and blood sampling; (2) carcass weight: weight of reindeer after removal of antlers and gastrointestinal tract; (3) foreleg weight: weight after skinning the left shoulder ventrally to the elbow and dissecting away extrinsic muscles at their insertions; (4) hindleg weight: after skinning the left haunch ventrally to the knee (but leaving the rump fat intact), dissecting away extrinsic muscles and separating intrinsic muscles and fat from the pelvis; (5) gastrointestinal weight: the weight of the gastrointestinal tract and its contents was calculated as (total weight − [carcass weight + antler weight]); (6) liver weight; (7) kidney weight: the combined weight of both kidneys after removal of perinatal fat. In addition, total weight of all sexed foetuses was measured.

A logistic curve was fitted to total weight data from −6 to 12 months of age, as described for crown-tail length. For all other parameters including total weight, sine waves were used to describe the relationship between weight and age for animals from birth onwards. A curve of the form $Y = (A - Be^{-kt})(1 + C\sin[2\pi t/12 - \Phi])$ was fitted by weighted least squares, where C = proportional fluctuation about the growth curve, and Φ = phase adjustment needed to align the theoretical sine wave with the annual reproductive cycle (considered to start on 15 November). This curve has an underlying growth component $(A - Be^{-kt})$ which is an equation for self-inhibiting growth. This is more appropriate than an equation for self-accelerating growth used elsewhere for reindeer and other cervids (Wood, Cowan & Nordan, 1962; McEwan, 1968; Bandy, Cowan & Wood, 1970) because points of inflection for changes in length and weight occur at, or soon after, birth (Leader-Williams & Ricketts, 1982a). An additional component $(C\sin[2\pi t/12 - \Phi])$ reflects an annual fluctuation whose amplitude is proportional to the weight of the first component. In this model, maximum weights occurred at $\Phi + 3$ months, and minimum weights at $\Phi + 9$ months relative to mid-November. For all parameters, the fitting of a sine wave significantly $(P < 0.001)$ reduced the residual sum of squares compared with that derived from fitting a standard growth curve alone (e.g. for total weight, $F_{2,46} = 28.75$ in males and $F_{2,75} = 9.75$ in females; for liver weight, $F_{2,42} = 40.62$ in males and $F_{2,69} = 38.75$ in females). Differences in growth rates, asymptotic weights, phase adjustments and proportional fluctuation were tested by fitting common parameters and performing an analysis of variance.

Total body weight increases logistically from early foetal life to 12 months of age (Figure 6.9), supporting use of sigmoid curves for describing growth from the early stages of foetal life to one year of age amongst cervids (Robbins & Robbins, 1979). In reindeer, growth rate does not differ between males and females and the point of inflection of the curves occurs at 2 months, when calves weigh 25–28 kg. Hence the points of inflection are not related to puberty (cf. Brody, 1945), which occurs at 4–8 months of age in males and at 16 months in females on South Georgia (Chapter 4). By 12 months of age, males are heavier than females (Table 6.2) but have achieved less of their asymptotic total body weight (Figure 6.8). Unlike skeletal parameters, growth rates of all body and organ weights (excepting kidney weight) are faster from birth onwards in females than in males (Table 6.2), supporting earlier work on caribou and other cervids (McEwan, 1968; Bandy *et al.*, 1970). However, males achieve heavier body and organ weights than females (Figures 6.9, 6.10, Table 6.2). Therefore, males continue to grow for longer than females (Figure 6.8), but each component represents a similar proportion of the total body weight in each sex (Figure 6.5).

Body and organ weights also show seasonal changes and these are considered in three groups: (1) body and leg weights; (2) gastrointestinal weight; (3) organ weights. Body and leg weights all fluctuate in phase and by the same amount within each sex, but the phasing and size of the fluctuations differ between males and females. In contrast, gastrointestinal weight and liver and kidney weights change in phase and by the same amounts in males and in females (Table 6.2). The differences in phasing show that organ weights change first in the annual cycle, with the maximum weights occurring in mid-summer and the minima in midwinter in both sexes (Figure 6.11). This is followed by changes in gastrointestinal weight and by changes in body and leg weights in males. Changes in body and leg weights in females occur last in the annual cycle. The phase changes in body and leg weights reflect the occurrence of an autumn loss in weight after the rut in males, and of a late winter and

Figure 6.9. Growth and cyclic fluctuations in mean total weight of male and female reindeer. Fitted curves for reindeer from −6 to 12 months of age are logistic (arrow indicates time of birth); growth of reindeer from birth onwards is described by a standard growth curve incorporating a sine wave. For clarity, the two curves are merged where the estimated values for each coincide (five months for males and seven months for females). Inset shows the amplitude and phasing of one annual cycle for each sex (from Leader-Williams & Ricketts, 1982a).

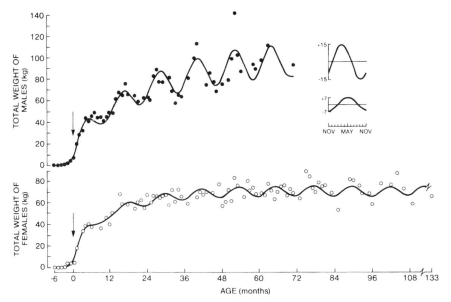

spring loss in females, during late pregnancy and early lactation (Figure 6.1). Furthermore, the fluctuations in body and leg weights are twice as great in males as in females (Figure 6.11).

Changes in body weight in *Rangifer* and other cervids are influenced by changes in gut weight (Mitchell, McCowan & Nicholson, 1976; Staaland, Jacobsen & White, 1979). On South Georgia, gastrointestinal weight forms a large proportion of body weight (Figure 6.5) and probably has a marked influence on changes in body weight, especially in males where the two weights change in phase. Furthermore, muscle tissue appears to change in weight over the annual cycle. Hence, the foreleg comprises mainly muscle and bone, yet shows similar changes to the hindleg (Table 6.2) which also carries a large depot of superficial body fat (Section 6.5). *Rangifer* appear adapted to sustain a negative nitrogen balance in winter and probably metabolise muscle when necessary (Reimers, Ringberg & Sørumgard, 1982; Syrjälä-Qvist & Salonen, 1983). However, much of the seasonal change in body weight of *Rangifer* arises from deposition and

Figure 6.10. Growth and cyclic fluctuations in mean liver weight of male and female reindeer. Growth is described by a standard growth curve incorporating a sine wave. Inset shows the similar amplitude and phasing of one annual cycle for both sexes (data from Leader-Williams & Ricketts, 1982a).

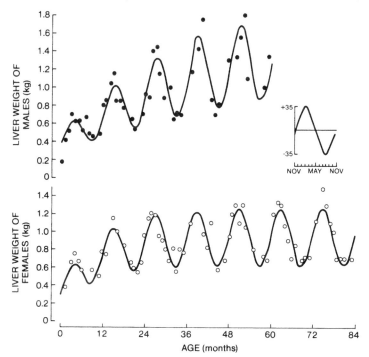

Table 6.2. *Growth and seasonal changes in body and organ weight of reindeer. The total weight up to 12 months includes foetal data: all other weights are from birth onwards. All weights are in kg, excepting kidney weight (g). Maximum weights occur at $\Phi+3$ months and minimum weights at $\Phi+9$ months relative to mid-November (from Leader-Williams & Ricketts, 1982a)*

Parameter and sex	Growth rate (k)	Asymptotic weight (A)	Phase adjustment (Φ)	Proportional fluctuation (C)	No. of weighted means
Total up to 12 months					
Male	{0.91}	45			19
Female		41			15
Total					
Male	0.037	103	1.0	15	51
Female	0.073	71	3.0	7	80
Carcass					
Male	0.031	75	1.0	15	51
Female	0.072	49	3.0	7	80
Foreleg					
Male	0.039	5.0	1.0	15	51
Female	0.083	3.0	3.0	7	78
Hindleg					
Male	0.044	8.0	1.0	15	51
Female	0.098	6.0	3.0	7	80
Gastrointestinal					
Male	0.052	28.0	{1.0}	{10}	51
Female	0.069	22.0			80
Liver					
Male	0.025	1.63	{−0.5}	{35}	47
Female	0.091	0.94			74
Kidney					
Male	{0.044}	217	{−0.5}	{36}	51
Female		182			79

mobilisation of fat (McEwan, 1968; Dauphiné, 1976; Reimers *et al.*, 1982).

Analysis of variance shows there is no difference in growth rate (k) for total weight of reindeer between -6 and 12 months of age ($F_{2,28} = 2.51$, $P > 0.10$), but that at 12 months of age males are heavier than females ($F_{1,30} = 11.84$, $P < 0.005$). In reindeer from birth and onwards, k is greater in females than in males for total weight ($F_{1,121} = 11.64$, $P < 0.001$), carcass weight ($F_{1,121} = 11.33$, $P < 0.005$), foreleg weight ($F_{1,119} = 10.66$, $P < 0.005$), hindleg weight ($F_{1,119} = 11.74$, $P < 0.001$), gastrointestinal weight ($F_{1,121} = 49.15$, $P < 0.001$), and liver weight ($F_{1,111} = 14.85$, $P < 0.001$); only for kidney weight do values of k not differ ($F_{1,120} = 3.43$, $P > 0.10$). However, all values of A are heavier ($P < 0.001$) in males than in females (e.g. for total weight, $F_{1,121} = 227.16$; hindleg weight, $F_{1,121} = 194.90$; gastrointestinal weight, $F_{1,121} = 123.09$; liver weight, $F_{1,11} = 140.13$).

Both the phasing (Φ) and proportional changes (C) differ between the sexes for all four body and leg weights (e.g. for total weight, $F_{1,121} = 5.25$, $P < 0.025$ and $F_{1,121} = 11.73$, $P < 0.001$ respectively). Furthermore, all four fluctuate in phase and by the same proportions (all $P > 0.10$) within each sex (e.g. for total weight, $F_{2,46} = 1.16$ in males and $F_{2,75} = 0.83$ in females), and common values are shown for all body and leg weights for males and for females in Table 6.2. The values of Φ and C for gastrointestinal weight are similar in males and females ($F_{2,111} = 1.58$, $P > 0.10$). Values of Φ are common for liver and kidney weights in males and females (all $P > 0.10$). However, values of C are the same for

Figure 6.11. Phasing and proportional fluctuation in various body and organ weights over one annual cycle in reindeer (from Leader-Williams & Ricketts, 1982a).

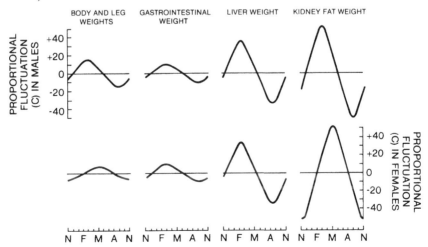

each sex for each of liver and kidney weights, but differ between parameters (e.g. for liver weight of females, $F_{2,75} = 3.27$ for Φ; $F_{1,73} = 9.52$, $P < 0.01$ for C).

6.5   Body condition

Fat reserves and blood chemistry are often measured as an index of population condition (Hanks, 1981). Continental reindeer and caribou live in a highly seasonal environment and show marked fluctuations in their fat reserves and in various parameters of their blood chemistry (e.g. McEwan & Whitehead, 1969; Hyvärinen, Helle, Väyrynen & Väyrynen, 1975; Dauphiné, 1976; Nieminen, 1980b; Timisjärvi, Nieminen & Saari, 1981; Reimers, 1983c; Nieminen & Laitinen, 1986). However, accumulation and loss of fat is even more marked in Spitzbergen reindeer (Figure

Figure 6.12. A female Spitzbergen reindeer, showing the rotund appearance of this sub-species, which accumulates far thicker layers of back fat than continental tundra and woodland sub-species of *Rangifer*. (Photograph by N.J.C. Tyler.)

6.12) (Krog *et al.*, 1976; Ringberg, 1979; Reimers *et al.*, 1982; Tyler, 1987a,b). On South Georgia I took one measure of blood chemistry (serum protein) and three measures of the fat reserves of reindeer. These reserves are sequentially mobilised during adverse periods, as follows: rump fat, kidney fat and bone marrow fat. They are laid down during favourable periods in the reverse order (Riney, 1955; Dauphiné, 1976; Hanks, 1981). I have not calculated the kidney fat index because of changes in kidney weight (Table 6.2), but instead I simply present data on weight of the fat surrounding the kidneys (cf. Dauphiné, 1975, 1976). In this section, I describe changes in fat reserves and serum protein with respect to age and season (Leader-Williams & Ricketts, 1982a). I also examine the effect of body condition on the fertility of females.

6.51 *Changes in fat reserves and serum protein*

Changes in fat reserves and serum protein were assessed by taking the following measurements, as described fully in Leader-Williams & Ricketts (1982a): (1) kidney fat weight: the combined weight of kidney fat from both sides, without trimming at the anterior and posterior ends of each kidney, was calculated by subtracting kidney weight from weight of the kidneys with attached fat; (2) rump fat depth: the maximum depth of rump fat was measured on the skinned left hindleg; (3) femur marrow fat: the % fat content of a sample of bone marrow was measured by the dehydration method of Neiland (1970); (4) total serum protein: venous blood collected after shooting was separated by centri-fugation, and total serum protein was measured using Biuret reagent.

Kidney fat weights and serum protein concentrations both increase with age and fluctuate by season, and sine waves incorporating growth curves were fitted to these parameters. However, femur marrow fat did not show any significant growth, and a simple sine wave was fitted to those data. Differences in growth rate, asymptote, phase adjustment and proportional fluctuation were tested by fitting common parameters and performing an analysis of variance. Rump fat also fluctuates seasonally, but is absent for at least part of the year, so a sine wave formula is not appropriate. Data for different year classes were thus combined into blocks of months in which there was either a high (mostly 100%) or a lower frequency of zero readings. Differences in frequencies between year classes were compared by chi-square test, pairs of mean fat depths by Student's *t* test, and groups of mean fat depths by analysis of variance.

The growth rate of kidney fat is more rapid in females than in males, but higher kidney fat weights are eventually attained in males (Figure 6.13, Table 6.3). However, femur marrow fat shows no increase with age

Table 6.3. *Growth and seasonal changes in condition of reindeer from birth onwards, reflected by kidney fat weight (g), femur marrow fat (%) and serum protein concentration (mg ml⁻¹). Maximum weights occur at Φ+3 months and minimum weights at Φ+9 months relative to mid-November (from Leader-Williams & Ricketts, 1982a)*

Parameter and sex	Growth rate (k)	Asymptotic value (A)	Phase adjustment (Φ)	Proportional fluctuation (C)	No. of weighted means
Kidney fat weight					
Male	0.010	89	1.0	{51}	51
Female	0.106	34	3.0		79
Femur marrow fat					
Male	—	51	2.0	{45}	51
Female	—	61	3.0		79
Serum protein concentration					
Male	{0.118}	{5.8}	{−0.5}	{16}	47
Female					67

and is generally higher in females. Serum protein concentrations, in contrast, show a similar pattern of growth in both sexes (Table 6.3). Kidney fat weight and femur marrow fat fluctuate by the same amount in both sexes (Figure 6.11, Table 6.3). However, changes in fat reserves are out of phase in males and females. In females, both fat reserves change in phase with body and leg weights. In males, kidney fat weight also changes in phase with body and leg weights, but the peaks and troughs in femur fat occur one month later. Serum protein concentrations fluctuate in phase and by the same proportion in males and females, and change in phase with liver and kidney weights.

Rump fat is not found in male or female calves until three months of age, is present in small amounts from four to seven months, and has disappeared by eight months (Figure 6.14). In yearling and adult males, rump fat occurs most frequently in the two months preceeding the rut, and the proportion of males with fat deposits then increases with age. Indeed, fat deposits are deeper in males over three years of age. In yearling females, rump fat is found from two months before the rut until the first half of the winter, during gestation. In females that had given birth, rump fat usually remains absent until after the following rut, and mostly disappears from pregnant females by the end of winter. There is no difference in peak fat depths between different age classes of females amongst either adults or adults, yearlings and calves (Figure 6.14).

Figure 6.13. Growth and cyclic fluctuations in mean kidney fat weight of male and female reindeer. Growth is described by a standard growth curve incorporating a sine wave. Inset shows the amplitude and phasing of one annual cycle for each sex (data from Leader-Williams & Ricketts, 1982a).

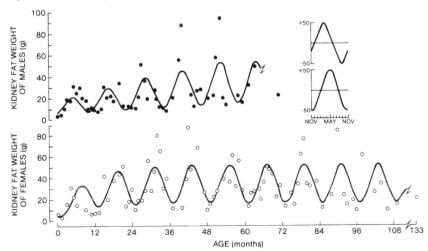

These results show a clear dimorphism between the sexes in accumulation and utilisation of their fat reserves. These changes, like those in body and leg weights, are closely related to the periods of major energetic demands associated with reproduction in each sex (Figure 6.1). The size and extent of the fat reserves of reindeer on South Georgia are similar to those of barren-ground and woodland caribou, and Norwegian and Finnish reindeer (Dauphiné, 1976; Parker, 1981; Reimers *et al.*, 1983; Nieminen & Laitinen, 1986) but far smaller and less dramatic than those seen in Spitzbergen reindeer (Reimers & Ringberg, 1983; Tyler, 1987a,b). In spite of this major difference between sub-species of *Rangifer*, the timing of the loss of fat in male and female reindeer is similar on South Georgia and Svalbard (Tyler, 1987a,b).

Analysis of variance shows there is no significant reduction in the fit of a sine wave which incorporated a growth curve, when compared to the fit of a sine wave without the growth component, to femur marrow fat %

Figure 6.14. Changes in rump fat depth of reindeer. Shaded blocks give the proportion of reindeer with no rump fat; circles are the mean ± SE of rump fat depth. (A) males over three years and adult females (excluding non-lactating females from November to April and non-pregnant females from May to November); (B) yearling and 2–3 year old males, or yearling females (excluding non-pregnant females as above); (C) calves. The rut is shown with * (from Leader-Williams & Ricketts, 1982a).

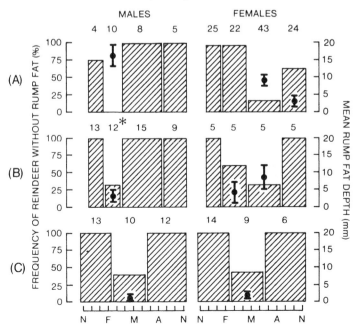

($F_{2,46} = 2.06$, $P > 0.10$): the growth component is therefore excluded for this parameter. There is a difference in values of both k and A for kidney fat weight between males and females ($F_{1,120} = 52.35$ and $F_{1,121} = 57.26$, both $P < 0.001$); values of A also differ for femur fat % ($F_{1,124} = 13.0$, $P < 0.001$). However, values of k and A are similar between the sexes for serum protein concentrations ($F_{1,104} = 0.07$ and $F_{1,105} = 0.31$, both $P > 0.10$) and common values are shown where appropriate in Table 6.3. The phasing of changes in kidney fat weights and femur fat % differ between the sexes (e.g. for kidney fat weight, $F_{1,120} = 4.90$, $P < 0.05$). However, changes in the parameters are in phase in females ($F_{1,74} = 0.49$, $P > 0.10$), but differ for males ($F_{1,46} = 5.27$, $P < 0.05$). Serum protein concentrations fluctuate both in phase and by the same proportion in each sex ($F_{2,104} = 1.04$, $P > 0.10$), and common values are shown in Table 6.3.

The frequency of male reindeer with rump fat before the rut increases with age ($\chi^2 = 9.34$, $df = 3$, $P < 0.05$). Furthermore, mean fat depths differ across all age classes ($F_{3,28} = 8.42$, $P < 0.001$), but do not differ ($F_{2,19} = 1.89$, $P > 0.10$) between calf, yearling and 2–3 year old males. By contrast, there is no difference (both $P > 0.10$) in peak fat depths across age classes in females amongst either adults ($F_{5,36} = 0.88$) or adults, yearlings and calves ($F_{7,49} = 1.47$). Thus data for yearling and 2–3 year-old males, and for those of adult females are combined in Figure 6.14 because of their similarity.

6.52 *Effects of weight and condition on fertility*

Fat reserves and body weight are both known to affect fertility of female *Rangifer* and other cervids (Mitchell & Brown, 1974; Dauphiné, 1976; Hamilton & Blaxter, 1980; Thomas, 1982; Reimers, 1983b; Albon, Mitchell, Huby & Brown, 1986). Because reindeer have high fecundity rates on South Georgia (Chapter 4), changes both in fat reserves and in body weights have been analysed using all females (with the exception of the analysis of rump fat shown in Figure 6.14). However, females of varying status were also examined for possible differences in condition and weight (Leader-Williams & Ricketts, 1982a). Nulliparous females have lower carcass and kidney fat weights during the winter than pregnant females due to calve at 2 and 3 years of age (Figure 6.15). In contrast, nulliparous females that have not calved at 2 years of age have heavier carcass weights than post-partum lactating females, though their kidney fat weights do not differ. It appears that female reindeer on South Georgia attain a weight and condition threshold before first conceiving as in other cervids (Hamilton & Blaxter, 1980; Reimers, 1983b; Albon *et al.*, 1986). However, if young females do not conceive, their body weight increases relative to females that have bred.

During the winter, non-pregnant parous females have heavier carcass weights than pregnant females, though their kidney fat weights do not differ. The high carcass weights of non-pregnant females is unexpected, but three of this sample were old females. In summer, however, non-lactating females have both heavier carcass and kidney fat weights than post-partum lactating females (Figure 6.15). Thus costs of lactation result in lower gains of weight and condition during summer for nursing females than for those with no calf, which in turn results in a decreased likelihood of conception at the subsequent rut (cf. Mitchell & Brown, 1974; Dauphiné, 1976). The overall relationship between body weight and fertility has been confirmed independently by comparison of the simple regression of luteal scars on age (Figure 4.11A) with a multiple regression of luteal scars on age and body weight. This comparison shows that heavier females are more fertile, as measured by numbers of luteal scars.

> Mean carcass (less conceptus, where appropriate) and kidney fat weights were calculated for all females of each year class collected in one of four seasonal age groups, as described in Leader-Williams & Ricketts (1982a): the first four months, and the fifth to seventh month of

Figure 6.15. Deviations from mean carcass and kidney fat weights of females of different reproductive status. Significant differences between deviations shown with * = $P < 0.05$ and ** = $P < 0.01$. Sample size is shown below columns (from Leader-Williams & Ricketts, 1982a).

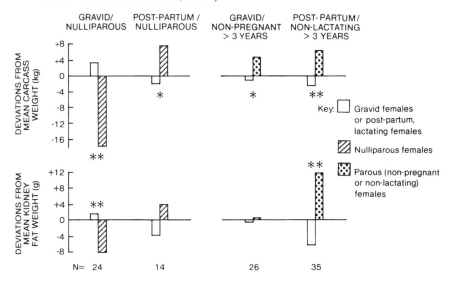

gestation; the first two months of lactation; the remainder of the summer until conception. Individual deviations from these means for females of different status were ranked and compared by the Mann-Whitney U test, using two-tailed probability values. In winter nulliparous females have lower (both $P < 0.01$) carcass ($z = 2.63$) and kidney fat weights ($z = 2.71$) than gravid females of the same ages. In summer, nulliparous females have heavier carcass weights ($U = 0$, $n_1 = 3$, $n_2 = 11$, $P < 0.02$), but their kidney fat weights do not differ ($U = 10$, $P > 0.10$) from post-partum, lactating females. In winter non-pregnant, parous females have heavier carcass weights ($z = 2.27$, $P < 0.05$) than gravid females, but their kidney fat weights do not differ ($z = 0.36$, $P > 0.10$). In summer, non-lactating females have both heavier carcass ($z = 3.25$, $P < 0.005$) and kidney fat weights ($z = 3.25$, $P < 0.005$) than post-partum, non-lactating females. The simple regression of luteal scars on age has been shown previously to be of the form $y = -0.81 + 0.93x$ (see Figure 4.11A). There is a significant reduction in residual variance ($F_{1,152} = 5.46$, $P < 0.05$) between this simple regression and a multiple regression of luteal scars on age and carcass weight. The regression is of the form $y = -2.02 + 0.89x_1 + 0.03x_2$, and confirms that weight affects fertility.

6.6 Growth, seasonality and sexual dimorphism

All temperate cervids undergo seasonal changes in body weight and fat reserves (moose: Franzmann, LeResche, Rausch & Oldmeyer, 1978; red deer: Mitchell *et al.*, 1976; wapiti: Flook, 1970; white-tailed deer: Moen, 1978; black-tailed deer: Bandy *et al.*, 1970; Anderson, Medin & Bowden, 1974; roe deer: Prior, 1968; Strandgaard, 1972b; fallow deer: Chapman & Chapman, 1975). There is now increasing evidence that these changes have an intrinsic basis and are adaptations to periods when forage availability is limited both in quality and quantity and when no plant growth occurs (Ryg, 1986). Thus voluntary reduction in the intake of food and of energy and nitrogen in captive deer fed *ad libitum* corresponds with the observed reduction in metabolic rate of deer in winter (McEwan, 1968; McEwan & Whitehead, 1970; Silver, Colovus, Holter & Hayes, 1969; Gasaway & Coady, 1974; Moen, 1978). In addition levels of circulating hormones, including growth hormone and somatomedin, cortisol, and thyroid hormones and prolactin show seasonal variations (Yousef, Cameron & Luick, 1971; Ringberg, Jabcobsen, Ryg & Krog, 1978; Ringberg, 1979; Ryg & Jacobsen, 1982a,b,c; Ryg, 1986). These intrinsic variations are likely to be influenced ultimately by external factors such as photoperiod, a decrease in which results, for example, in the seasonal inappetance in red deer (Pollock, 1974; Kay & Staines, 1981; Ryg, 1986).

On South Georgia, both the voluntary reduction in food intake of reindeer and the natural limitation of forage quality and quantity combine to result in changes in rumen weight and in available ruminal nitrogen that are common to both sexes (Chapter 5; Figure 6.16). The autumn switch to protein-poor tussock grass, and the spring change to more nutritious forage species on South Georgia is paralleled by similar changes amongst temperate cervids in the Northern Hemisphere (see references in Chapter 5). Experimental studies on glucose metabolism, and on renal excretion and recycling of urea (Luick, Person, Cameron & White, 1973; Hove & Jacobsen, 1975; Wales, Milligan & McEwan, 1975; Bjarghov *et al.*, 1976) suggest that the factor which most affects liver and kidney weights is dietary quality. Thus changes in the weights of these organs, and in serum protein which is synthesised by the liver, are not related to sex differences in reproductive phenology, but rather to changes in dietary diversity and available nitrogen.

Superimposed upon these changes are the effects of the sex difference in reproductive phenology, such that the greatest energetic demands occur for males at the rut and for females in the last third of pregnancy and early lactation (Verme, 1965; Moen, 1978). Thus there is a clear difference in phasing and in fluctuation between male and female reindeer on South Georgia for changes in body and leg weights and in fat reserves. All these measurements, including rump fat depth, are at a

Figure 6.16. Differences in the timing and size of proportional fluctuations in resources available (measured as total ruminal N) and in the energetic costs of reproduction (as measured by changes in body weights) for each sex.

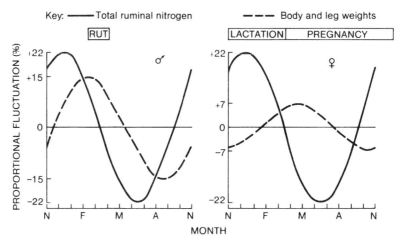

peak in males before the rut and begin to decrease during it, whereas body weights and fat reserves in females decline to a minimum around the time of calving and do not peak until late autumn. This result is in agreement with sex differences seen in other temperate cervids (see references cited above).

The interest of our analysis lies in the contrast between parameters that show simple seasonal changes in each sex (Figure 6.16). The feeding ecology of cervids may differ in each sex, but the difference between males and females tends not to be marked (Chapter 5; Clutton-Brock *et al.*, 1982). The resources available, here measured as ruminal nitrogen, are therefore similar or only slightly different for each sex. However, superimposed on this are marked differences in many aspects of body growth, and weight and fat loss patterns of males and females (Figure 6.16). This arises because the chances of reproductive success amongst males of a strongly polygynous species of cervid such as *Rangifer* are enhanced by the attainment of large body size and antlers (Clutton-Brock *et al.*, 1980, 1982). The costs to each sex, in terms of mortality, are considered in the next chapter.

6.7   Summary

1. The antlers of male reindeer are more complex and heavier, both absolutely and relative to body weight, than those of females. Males also break their antlers more frequently than females.
2. Growth in length and weight from early foetal life until one year of age is logistic. A sex difference in skeletal size becomes apparent only after one year of age although by then males are already significantly heavier than females.
3. Organ and gastrointestinal tract weights change first in the annual cycle. These changes show little or no dimorphism in timing or proportion between the sexes.
4. In contrast, changes in body and leg weights and in fat reserves are related to reproductive activity. In males they decline after the rut whereas in females they reach a minimum around parturition.
5. Although both sexes probably metabolise muscle protein as well as fat during winter, male reindeer lose twice as much of their body tissue during the annual cycle as females.
6. These results indicate that the cost of reproduction is greater to males than to females and this reflects their different reproductive strategies.

7

Causes of mortality

> When a species, owing to highly favourable circumstances, increases
> inordinately in numbers in a small tract, epidemics . . . often ensue.
> Charles Darwin (1859)

7.1 Introduction

Disease and predation, together with food shortage, have long
been recognised as important density-dependent factors that limit mammal numbers (Lack, 1954; Anderson, 1982; Fowler, 1987). The interplay
of each factor in contributing towards death is complicated, and difficult
to study in the field. Yet changes in the rate of mortality probably have a
greater influence upon population dynamics of ungulates than changes in
fecundity rate (Caughley, 1970; Sinclair, 1977; McCullough, 1979;
Clutton-Brock, Major & Guinness, 1985a). Changes in the relative
importance of the causes of mortality clearly underlie any actual change
in mortality rate. Thus the causes of mortality should be documented in
any study of population dynamics.

Several factors may affect the patterns and causes of mortality
observed on South Georgia. First, there are no native or introduced
mammalian predators on the island. In contrast, predation is a major
cause of mortality in many populations in the Northern Hemisphere
(Bergerud, 1974c, 1980a, 1983; Gauthier & Theberge, 1986). Secondly,
there has been little hunting of reindeer on South Georgia, especially in
the past three or four decades (Chapter 3). In contrast, many northern
populations of *Rangifer* are now limited by hunting (Bergerud, 1974c;
Miller, 1983; Bergerud *et al.*, 1984). Thirdly, reindeer herds on South
Georgia are founded on the small numbers originally introduced (Chapter 3). Inbreeding can have deleterious effects upon ungulates and may
influence patterns of mortality (Ralls, Brugger & Ballou, 1979; Ballou &
Ralls, 1982; Loudon, 1985). Fourthly, reindeer now occur at high

densities and in unusual environmental conditions compared with northern populations.

In this chapter, I examine in turn the patterns (7.2), rates and general causes of mortality (7.3), the parasite burden (7.4) and the most common disease (7.5) of reindeer on South Georgia. Finally, I compare patterns of mortality amongst different populations of reindeer and caribou (7.6). Throughout the chapter I relate findings to differences already described between Northern and Southern Hemisphere populations. I use data in this chapter to question whether differences in mortality patterns in males and females are determined by differences in susceptibility to diseases, or whether they are related to other factors.

> Data are derived from composition counts, a sample of 146 carcasses resulting from natural deaths found in all three herds, and also from post-mortem examinations of the shot sample (Leader-Williams, 1980b). Composition counts were carried out to determine ratios of males and females in different segments of the population. The time of death of 37 calf carcasses was determined from skeletal measurements (femur, hindfoot, mandible and diastema lengths) which were compared with growth curves derived from the shot sample (Chapter 6). A total of 14 adult and yearling reindeer carcasses were sexed directly, and a further 85 by the disparity in antler size and morphology. The remaining 10 carcasses were sexed only after they had been assigned an age, when it was possible to determine their sex from differences in skeletal measurements. The time of death of adult and yearling reindeer carcasses was assessed from sex differences in the antler cycles (Chapter 4). The presence of broken bones and dental abnormalities was noted in all carcasses and a post-mortem examination was performed where possible on flesh-covered carcasses. All pathological conditions found in the shot sample were recorded.

7.2   Patterns of mortality

Reindeer and caribou show a high degree of sexual dimorphism in body size (Chapter 6). Such dimorphism is generally associated with a mortality differential between the sexes (e.g. Clutton-Brock *et al.*, 1982; Clutton-Brock, Albon & Guinness, 1985b). In this section I examine differences in sex ratios and in the timing of deaths between males and females. I compare patterns of mortality on South Georgia with those on Svalbard, which also lacks predators.

7.21   Mortality differential

Comparison of foetal sex ratios (Leader-Williams, 1980b) and sex ratios of reindeer of different ages obtained during herd composition counts (Table 7.1) shows a mortality differential on South Georgia. The ratios of males to females decline progressively with increasing age (Figure 7.1). The mortality differential is also evident from the age structure of the shot sample (Figure 3.12) and of the carcasses arising from natural deaths (Leader-Williams, 1980b). The longevity and survival of males and females differs (Figure 7.2). A similar pattern is seen in Spitzbergen reindeer, even though reindeer on Svalbard live longer than on South Georgia (de Bie, 1976; Reimers, 1983a), and in all other *Rangifer* populations, whether or not they are subjected to predation or hunting (e.g. Kelsall, 1968; Parker, 1972; Reimers, 1975; Skogland, 1985).

> Sex ratio data were derived as described in Leader-Williams (1980b): (1) foetuses were collected from shot females; (2) calves of 6 months and 1 year and (3) adults and yearlings were counted during composition counts. Data for 1975 and 1976 were combined. A chi-square test shows there is a progressive decline in males across all age classes from foetuses to adult reindeer ($\chi^2 = 45.75$, $df = 3$, $P < 0.001$). Survival curves (Figure 7.2) are explained below.

7.22   Time of death

The sample of carcasses located from all three herds comprised 25% calves, 8% yearlings and 67% adults. Amongst calves the sex of carcasses could not be distinguished. However, there are two distinct

Table 7.1 *Composition counts in the Barff herd (data from Leader-Williams, 1980b)*

Date	N	Calves/100 adults and yearlings	Calves/100 adult and yearling females	Calves/100 adult females	Adult and yearling males/100 adult and yearling females	Adult males/100 adult females
April 1974	686	36.1				
March 1975	529	34.6	47.8		38.4	
March 1976	739	35.8	51.5	63.9	43.5	29.5
October 1974	566		25.5			
October 1975	567		44.8			

periods in which deaths occur (Figure 7.3). Half the carcasses located had died within one month of birth, and several were so small that they had probably succumbed soon after birth. The remaining calf carcasses form a less distinct grouping whose time of death varied between March and the following November, with a peak during winter months. Therefore, calf deaths on South Georgia are evenly distributed between the perinatal period and the first winter, but rarely occur during the first summer, as in red deer (Guinness, Clutton-Brock & Albon, 1978b). However, this contrasts with carcass finds of reindeer on Svalbard (Reimers, 1983a). There 95% of calves die during winter months and only 5% of deaths arise from the perinatal period (Figure 7.4).

Figure 7.1. Sex ratios of reindeer of various ages on South Georgia. Sample sizes shown as appropriate (data from Leader-Williams, 1980b and Table 7.1).

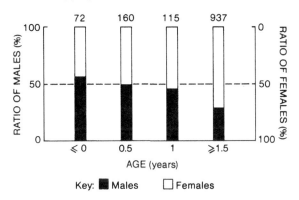

Figure 7.2. Survivorship of male and female reindeer on South Georgia and Svalbard (data from Table 7.3 and Reimers, 1983a).

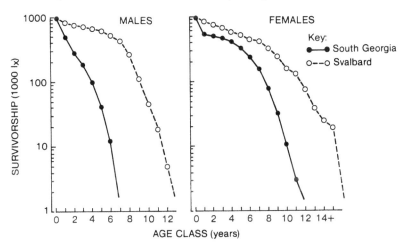

Differences in the antler cycles (Figure 4.18) allow the seasons when males and females die to be identified (Figure 7.5). Most adult and yearling males (68%) die in early winter, and a further 20% die in late winter or early spring. Most females (67%) die later in the winter than males and a further 12% die in early spring. Few female deaths occur in early winter, and there is little mortality of both adult and yearling males (12%) and females (9%) over summer months. Reindeer on Svalbard show a less exaggerated sex difference than on South Georgia, and most males die with cast antlers later in the winter (Figure 7.6).

Chi-square tests show there is no difference ($\chi^2 = 8.24$, $df = 6$, $P > 0.10$) between the distribution of carcasses in perinatal, winter calf, yearling and adult groups in three herds on South Georgia, and all data are combined. The distribution of carcasses in these groups differed ($\chi^2 = 36.78$, $df = 3$, $P < 0.001$) between South Georgia and Svalbard.

Figure 7.3. Time of death of 37 calf carcasses from all herds, assessed from skeletal measurements (redrawn from Leader-Williams, 1980b).

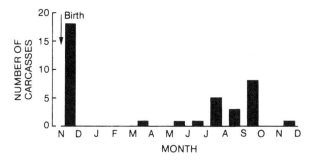

Figure 7.4. Frequency of occurrence of carcass finds in different age groups on South Georgia and on Svalbard. *** signify a difference of $P < 0.001$ across groups (data from Leader-Williams, 1980b; Reimers, 1983a).

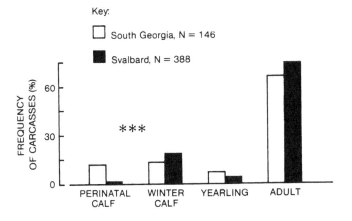

Figure 7.5. Time of death of 109 adult and yearling reindeer carcasses from all herds, assessed from state of antlers (data from Leader-Williams, 1980b).

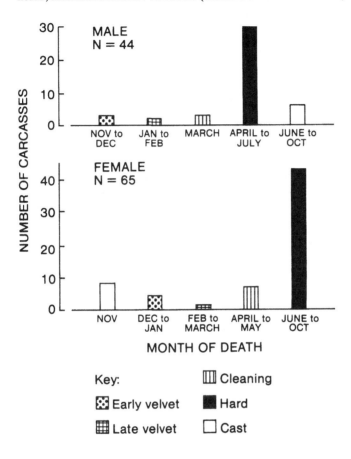

Figure 7.6. Differences in timing of death of adult and yearling reindeer on South Georgia and on Svalbard, assessed from antlers. More males (*** signify a difference of $P < 0.001$) die in early winter on South Georgia, but timing of female deaths does not differ.

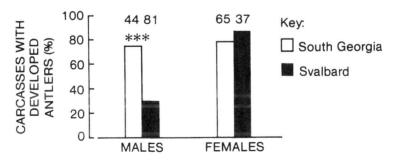

There is no difference between the distribution of adult and yearling males ($\chi^2 = 9.91$, $df = 8$, $P > 0.10$) or females ($\chi^2 = 15.12$, $df = 8$, $P > 0.05$) in different antler categories in the three herds, and data from all herds are combined. More males on South Georgia die in early winter than on Svalbard ($\chi^2 = 18.57$, $P < 0.001$), but the timing of female deaths does not differ ($\chi^2 = 1.00$, $P > 0.10$).

7.3   Rates and causes of mortality

7.31   Calves

The rates of mortality amongst perinatal calves were estimated from the difference between pregnancy rates and late-summer calf counts (Table 7.2). Rates of perinatal mortality were between 15 and 30% amongst the three herds on South Georgia for the November 1975 calf crop. I could not investigate the causes of perinatal mortality in detail as few carcasses were fresh enough for examination. However, the rates of perinatal mortality on South Georgia differ from many northern populations, because herds exposed to predation usually lose 50% or more of their calves (Bergerud, 1980a). Most perinatal deaths in such herds arise from predation, and calf abandonment and various physiological and pathological disorders are less important causes of mortality (Bergerud, 1971; Miller & Broughton, 1974; Nordkvist, 1980; Miller, Broughton & Gunn, 1983). Some domestic reindeer and wild *Rangifer* herds largely lack predators, however, and calf abandonment, hypothermia, accidents and infections have been cited as the main causes of perinatal death (Baskin, 1970, 1983; Clausen *et al.*, 1980; Thing & Clausen, 1980; Eloranta & Nieminen, 1986). Similar causes are likely to result in perinatal deaths on South Georgia.

The next phase of calf mortality on South Georgia occurs during their first winter (Figure 7.3), but direct evaluation of mortality rates was not possible from differences in late-summer and late-winter calf counts

Table 7.2. *Estimation of perinatal mortality in November 1975 from pregnancy rates and late summer calf counts (from Leader-Williams, 1980b)*

Estimate	Barff herd	Busen herd	Royal Bay herd
A Pregnancy rate	0.90	0.93	0.91
B Calves/adult females	0.64	0.79	0.77
C % calf loss, $[(A-B)/A]$ 100	29	15	15

(Table 7.1) because of possible between-winter differences in mortality of females. No fresh carcasses arising from winter deaths were found for examination and data on their body condition are therefore lacking. However, calves have only small reserves of carcass fat at the start of winter (Chapter 6) and this is probably a major contributory cause of calf mortality. In addition, accidents arising either from cliff falls or avalanches commonly result in death of calves in winter (Figure 7.7). Indeed, a snowslip wiped out the first reindeer herd introduced to the Busen area (Chapter 3). A similar accident, involving six calves and a yearling female, was recorded from the Royal Bay herd in winter 1975.

7.32   Adults and yearlings

Age-specific mortality rates amongst adult and yearling reindeer of both sexes have been estimated from life tables (Table 7.3). These were calculated from the age structure of shot reindeer, after correcting for bias in the collection of calves and yearlings. Estimates of mortality for reindeer on South Georgia (Table 7.3, Figure 7.8) follow the general pattern for *Rangifer* and other ungulates (Caughley, 1966, 1970; Parker, 1972; Peterson, 1977; Sinclair, 1977; Reimers, 1983a). Mortality of yearlings and adults increases progressively with age, and differs between

Figure 7.7. Frequency of carcasses in different age groups killed as a result of cliff falls and avalanches. *** signify a difference of $P < 0.001$ between frequency of accidental deaths in different age groups. Sample sizes shown above columns.

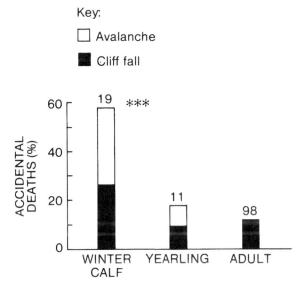

Table 7.3. *Life tables for male and female reindeer from the Barff herd*

Age (years)	Males						Females					
	N	Adjusted frequency (f_x)	Survival ($1000l_x$)	Deaths ($1000d_x$)	Mortality ($1000q_x$)	Expectation of life (e_x)	N	Adjusted frequency (f_x)	Survival ($1000l_x$)	Deaths ($1000d_x$)	Mortality ($1000q_x$)	Expectation of life (e_x)
0–1	78	73	1000	473	473	1.69	61	61	1000	498	498	3.24
1–2	40	38.5	527	219	416	1.76	33	30.6	502	17	34	4.97
2–3	19	22.5	308	109	353		32	29.6	485	18	37	
3–4	14	14.5	199	94	472		23	28.5	467	46	99	
4–5	10	7.7	105	60	571		21	25.7	421	75	178	
5–6	5	3.3	45	31	689		28	21.1	346	97	280	
6–7	0	1	14				15	15.2	249	97	390	
7–8							9	9.3	152	73	480	
8–9							5	4.8	79	46	582	
9–10							1	2.0	33	22	667	
10–11							0	0.7	11	8	727	
11–12							1	0.2	3			

males and females. A summary statistic can often convey better the impressions gained from life tables (cf. Caughley, 1970). This has been achieved by calculating the expectation of life of reindeer from life tables (Table 7.3). The difference in life expectancy of males and females confirms the mortality differential between the sexes.

Age-specific survival and mortality rates were calculated from numbers of reindeer shot in each age class. Shot reindeer over 2 years were assumed to have been collected randomly. A stationary age distribution was assumed because there was no difference in age structure of samples shot in 1974 and 1975 (males: $\chi^2 = 4.69$, $df = 3$; females: $\chi^2 = 0.54$, $df = 9$; both $P > 0.10$). There was probably a bias in the collection of calves and yearlings, so age structures were modified using ratios of calves and yearlings derived from composition counts, as described in Leader-Williams (1980b). Life tables were constructed in the standard manner. Age frequencies were smoothed using the probit model, and assuming a stationary age distribution (Caughley, 1977). Parameter estimates were obtained by maximum likelihood using the statistical package Genstat. For males, the probit model had an intercept = 6.3; a slope = -0.53; and a deviance = 5.01, $df = 2$, $P > 0.05$. For females, intercept = 8.0; slope = -0.50; deviance = 8.30, $df = 8$, $P > 0.10$.

Sex differences in the timing of mortality correspond with periods of loss of fat reserves and body weight, after the rut in males and during late pregnancy and around calving in females (Chapter 6). The metabolic demands of late pregancy and early lactation are high (Moen, 1973). In

Figure 7.8. Sampled and adjusted frequency in each age cohort, and age-specific mortality of female reindeer on South Georgia (data from Table 7.3).

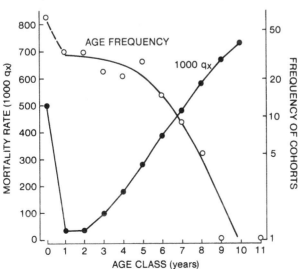

addition, forage has been restricted in quality and quantity (Chapter 5) at the time when females die. Thus female carcasses in which fat reserves could be measured had little femur marrow fat (10 ± 0.5%; $N = 5$), which is suggestive of death by starvation (Nieminen & Laitinen, 1986). Males dying after the rut do so without suffering the temporal restriction of forage during winter months. However, males lose body condition rapidly during the rut, and they lose more weight and condition than do females later in the annual cycle (Figure 6.11). Furthermore, males do not recover any fat reserves upon which to draw throughout the winter (Figure 6.14). On Svalbard, the pattern of fat loss had a similar timing to that on South Georgia (Tyler, 1987b) and males and females mostly die from starvation during winter months (Reimers, 1982, 1983a).

Accidental deaths occur more frequently on South Georgia than on Svalbard (Figure 7.9; Tyler, 1987a). Entanglement of antlers in wire and fishing nets, and interlocking of antlers by males whilst fighting at the rut cause a small number of accidental deaths on both islands. However, many more reindeer on South Georgia die in avalanches and from cliff falls (Figure 7.10). Moreover, this is certainly an underestimate, because carcass finds only result from falls that occur in inland locations and from the few sea-cliffs with sufficient foreshore beneath them for carcasses not to be washed away (Figure 7.10B). One such 200 m stretch of beach provided 31% of the cliff-fall carcasses in one search. Most (85%) cliff-fall deaths occur in winter months. Males and females of all age

Figure 7.9. Numbers of reindeer carcasses on South Georgia and Svalbard resulting from accidental deaths. *** signify differences of $P < 0.001$ between islands. The sample from both islands includes adults, yearlings and winter calves (data from Leader-Williams, 1980b; Reimers, 1983a).

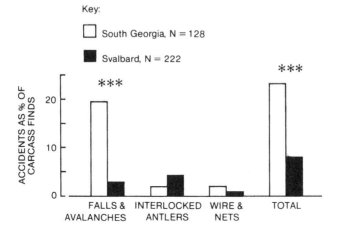

classes were equally susceptible to cliff falls, apart from a high proportion of winter calves (Figure 7.7). Animals that had survived falls were in poor condition due to loss of mobility and secondary infection arising from broken bones.

Cliff falls have increased in importance as a cause of death since the 1950s. Although several carcasses were then found which lay in places suggestive of cliff falls, a total of *c.* 50 carcasses were located in 14 days' fieldwork on the Barff Peninsula compared with my findings of 63

Figure 7.10. During winter tussock grass remains exposed on steep hillside cliffs (A), and many reindeer fall to their death (B).

carcasses in 118 days of summer fieldwork, including organised searches (Bonner, 1958; Leader-Williams, 1980b). This reflects both an overall decline in numbers (Figure 3.9), and a different pattern of mortality which now results in more cliff falls (and fewer carcass finds). Underlying this change is depletion of the most accessible tussock grassland due to overgrazing (see Figure 9.3), such that reindeer now have to venture to steeper and more dangerous cliffs to forage in winter (Figure 7.10A). Therefore, cliff-fall deaths probably occur as a result of food limitation in winter.

> Chi-square tests show that more accidental deaths occur on South Georgia than on Svalbard ($\chi^2 = 16.06$, $P < 0.001$), and that the major difference lies in numbers of cliff-fall deaths ($\chi^2 = 26.22$, $P < 0.001$). More calves and yearlings ($\chi^2 = 14.12$, $P < 0.001$) die from cliff falls and avalanches than do adult reindeer but there is no difference ($\chi^2 = 0.20$, $P > 0.10$) in numbers of males and females found dead from cliff falls. Many more carcasses per day of fieldwork were found by Bonner (1958) than during the present study ($\chi^2 = 35.76$, $P < 0.001$).

Parturition difficulties appear to account for 12% of female deaths, but few pathological conditions are seen on South Georgia (Bonner, 1958; Leader-Williams, 1980b). No diseases were likely to be primary causes of mortality, but instead were secondary causes that contribute to the ultimate demise of the animal. None was found more commonly in one sex than the other, apart from obvious examples like mastitis. I found only one or two cases of nephritis, hydronephrosis, liver abscess, mastitis or pneumonia. Generalised pyaemia was seen in six reindeer and the original locus of infection appeared to have been an interdigital abscess. Active peritonitis was seen in three reindeer and small areas of organised peritonitis, mainly around the abomasum, in 15 more. The only common diseases of reindeer on South Georgia are considered in more detail below.

7.4   Parasitism

The parasite fauna of ungulates is often geographically, rather than zoologically, characterised (Dunn, 1968). Two factors may determine the nature and importance of the parasite burden of reindeer on South Georgia (Leader-Williams, 1980c). Firstly, few potential intermediate hosts and no other ungulates occur on the island, possibly resulting in a parasite fauna that is poor in species. Secondly, reindeer live at high densities and continuously occupy their range without migrating, possibly leading to the development of heavy parasite burdens. Against this background I compare the parasite fauna on South Georgia with that of Northern Hemisphere reindeer and caribou and discuss the possible importance of those found as causes of mortality on South Georgia.

7.41 Warble flies

Neither the skin warbles, *Oedemagena tarandi*, nor throat warbles, *Cephenomyia trompe*, parasitise reindeer on South Georgia (Olstad, 1930; Bonner, 1958; Leader-Williams, 1980c). In contrast, most continental populations of *Rangifer* in the Northern Hemisphere are heavily (>80%) infested with both skin and throat warbles (Skjenneberg & Slagsvold, 1968; Bergerud, 1971; Nordkvist, 1971; Kelsall, 1975; Helle, 1980). Therefore, warble larvae were probably introduced to South Georgia, as happened when Norwegian reindeer were introduced to Greenland in 1952 (Rosing, 1956). However, warble larvae emerge from their host in the northern summer (May and June) and pupate in the ground for 33 to 58 days (Nordkvist, 1967). Because the seasons were reversed during transportation to South Georgia, pupation would have been unsuccessful on the snow-covered ground during the austral mid-winter. Any warble flies introduced are unlikely to have survived their first winter and reversal of seasons acted fortuitously as a quarantine that prevented their establishment on South Georgia.

7.42 Parasites with indirect life cycles

South Georgia lacks most specific intermediate hosts necessary for reindeer parasites that complete an indirect life cycle, particularly terrestrial molluscs. Infestation of reindeer by trematodes (liver flukes) or by protostrongylid nematodes is therefore not possible. Amongst the latter *Elaphostrongylus rangiferi* in Eurasia and *Pneumostrongylus tenuis* in North America can cause severe neurological symptoms and death in reindeer (Roneus & Nordkvist, 1962; Anderson, 1971; Lankester, Crichton & Timmermann, 1976; Halvorsen, Andersen, Skorping & Lorentzen, 1980). Of parasites with indirect life cycles, only *Moniezia (Blancharieza) benedeni* was found in the small intestine of c. 15% of reindeer on South Georgia. As the 18 species of native oribatid mites do not include any of the previously recorded intermediate hosts of ungulate cestodes (Gressitt, 1970; Sengbusch, 1977), *Moniezia* must have adapted to a new intermediate host (cf. Dunn, 1978).

7.43 Parasites with direct life cycles

The only parasites found commonly in reindeer on South Georgia are nematodes that complete a direct life cycle. As no other ungulates have shared the range of the reindeer or used it previously, these parasites must have been brought with the original stocks of reindeer from Norway.

> Data are derived from faecal samples and searches of gastrointestinal and respiratory tracts, as described by Leader-Williams (1980c). Quantitative counts were made on 94 faecal samples using the improved modification of the McMaster method (Anon., 1971) to determine numbers of strongylate nematode eggs. Faecal samples were cultured and infective (L_3) larvae collected using the Baermann apparatus (Anon., 1971). These larvae and adult nematodes found in searches of gastrointestinal tracts were preserved for identification.

Lungworms of the genus *Dictyocaulus* affect reindeer and caribou in the Northern Hemisphere, including in Norway (Skjenneberg & Slagsvold, 1968; Kummeneje, 1977). However, no signs of this parasite were seen on South Georgia.

In contrast, gastrointestinal nematodes are present in most reindeer. Two types of nematode eggs are found in faeces. The most common are typical ovoid strongylate eggs of 90 to 120 µm in length, laid by the abomasal parasite, *Ostertagia grühneri*, which was found in all reindeer. Eggs of a second type are larger (250 µm in length) and are laid by the small intestinal parasite, *Nematodirella longispiculata*, which was found in less in than 10% of reindeer. In addition, *Skjrabinema* species (probably *S. tarandi*) were recovered from the colon and caecum of less than 10% of reindeer, but they lay their eggs in the perineal region (Dunn, 1978). These species of nematode have all been reported from *Rangifer* previously (Hellesnes, 1935; Skrjabin *et al.*, 1954; Kelsall, 1968; Bergerud, 1971; Pryadko, 1976; Thing & Clausen, 1980; Bye & Halvorsen, 1983). Surprisingly, there appear to be no records of *O. grühneri* or *N. longispiculata* from reindeer on the Norwegian mainland, but *O. grühneri* occurs on Svalbard (Skjenneberg & Slagsvold, 1968; Bye & Halvorsen, 1983). The few parasites found on South Georgia (Table 7.4) conform to the prediction of a species-poor fauna due to the island's unusual environment, isolation from other ungulates and lack of intermediate hosts.

The only species of likely pathogenic significance on South Georgia is *O. grühneri*. Whilst ostertagiasis is an important parasitic disease of cattle and sheep (Dunn, 1978) it has not been recorded in reindeer or caribou, but might well occur (Rehbinder & Christennson, 1977). Faecal egg counts provide an indication of the importance of abomasal nematodes (Figure 7.11). Counts of *O. grühneri* eggs are always less than 600

Table 7.4. *Numbers of species and genera of gastrointestinal helminths found in different reindeer and caribou populations*

Locality	Sub-species or type	Abomasal nematodes		Small intestinal nematodes		cestodes		Large intestinal nematodes		Source
		Species	Genera	Species	Genera	Species	Genera	Species	Genera	
South Georgia	Introduced reindeer	1	1	1	1	1	1	1	1	This study
Svalbard	Spitzbergen reindeer	6	4			2	2			Bye & Halvorsen (1983); Bye (1985)
Canada – Newfoundland	Woodland caribou	2	1	2	2					Bergerud (1971)
Canada – NWT	Barren-ground caribou	1	1	1	1	3	3			Kelsall (1968)
Norway	Domesticated reindeer		3	3	3					Skjenneberg & Slagsvold (1968)
Sweden	Domesticated reindeer	>7	2	2	2	1	1			Rehbinder & Christennson (1977)
USSR	Domesticated reindeer	14	3	5	3	5	2			Skrjabin et al. (1954); Shalaeva (1975)

per gram in adults and yearlings, and do not differ between males and females. Data from each sex are therefore combined. Counts show a strong seasonal pattern and are high in spring and summer, and low or negative during winter. Thus, adult nematode populations are small in winter, as the parasite undergoes hyperbiosis in the abomasum (cf. Michel, 1969).

This pattern shows obvious epidemiological similarities with ostertagiasis in sheep. The rise in egg output (Figure 7.11) is analogous with the spring rise in ewes which provides the main source of infective larvae for lambs. Indeed, egg counts in reindeer calves are higher than in adults and yearlings, but do not attain the range seen in acute ostertagiasis in sheep (Dunn, 1978). As gastritis and diarrhoea do not occur in reindeer calves, ostertagiasis is unlikely to be a primary cause of mortality of reindeer on South Georgia. Instead, the parasites most probably lower growth rate and body condition of calves by reducing digestive efficiency and appetite in summer (cf. Sykes, 1978) and thereby contribute to death of calves in winter. The lack of heavy parasite burdens amongst the high density and non-migratory herds on South Georgia is contrary to expectations, and in direct contrast to Svalbard where abomasal parasites play a major role in the mortality of Spitzbergen reindeer (Bye & Halvorsen, 1983).

Mann-Whitney U tests show that strongylate egg counts from adult and yearling males do not differ from those of females (November 1974: $U = 33$, $n_1 = 8$, $n_2 = 12$; January: $U = 11$, $n_1 = 3$, $n_2 = 6$; August: $U = 11$, $n_1 = 3$, $n_2 = 7$; November 1975: $U = 25.5$, $n_1 = 7$, $n_2 = 9$; all $P > 0.10$).

Figure 7.11. Changes in strongylate nematode egg counts in faecal samples. Solid line joins mean ± SE of counts for 80 adult and yearling reindeer, whilst broken line joins mean ± SE of counts for 14 calves (redrawn from Leader-Williams, 1980c).

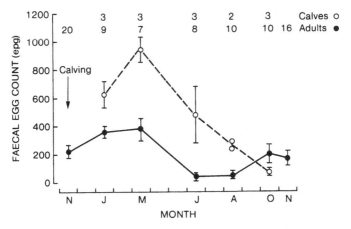

Kruskal-Wallis ANOVA shows that egg counts for all adults and yearlings combined ($N = 80$) differ between months ($H = 27.10$, $df = 6$, $P < 0.001$). This difference holds also for adult females ($N = 42$) only ($H = 16.53$, $P < 0.02$), but sample sizes for males and yearlings alone are too small to compare. Mann-Whitney U tests show that egg counts of calves are higher than those of adult and yearling reindeer (March: $U = 0$, $n_1 = 7$, $n_2 = 7$, $P < 0.02$; June: $U = 0$, $n_1 = 3$, $n_2 = 8$, $P < 0.02$).

7.5 Dental disease

Tooth wear and dental abnormalities may be an important cause of mortality amongst cervids (Spitzbergen reindeer: de Bie, 1976; Tyler, 1986; barren-ground caribou: Miller, 1974; red deer: Lowe, 1969; Mitchell, McCowan & Parrish, 1973; wapiti: Flook, 1970; fallow deer: Chapman & Chapman, 1975). Dental abnormalities and mandibular disease are common in reindeer on South Georgia (Leader-Williams, 1980a). In this section I discuss their prevalence and pathology, and assess the contribution of mandibular disease to mortality (Leader-Williams, 1982).

7.51 Prevalence

Dental abnormalities, many not serious, and mandibular disease have often been described in *Rangifer* (Banfield, 1954, 1961; Miller & Tessier, 1971; Miller, Cawley, Choquette & Broughton, 1975; Doerr & Dieterich, 1979). Here I concentrate only upon the more important damage to molar teeth and upon disease of the mandible. Molar damage includes lost, split or broken mandibular premolars and molars, whether or not they are accompanied by a mandibular swelling (Figure 7.12). Mandibular swellings are hard, often irregular lesions, sometimes impacted with vegetation, and always associated with molariform teeth.

No damage occurs amongst deciduous molar teeth, nor amongst any calves. However, molar damage occurs amongst 11% of adult and yearling reindeer in the Barff herd, and males and females are similarly affected. Molar damage is more common in animals with fully erupted teeth than in animals under 3 years of age with erupting teeth (Figure 7.13). m1, the first permanent molariform tooth to be established, and p4, the last to erupt, are most commonly affected (Figure 7.14). Mandibular swellings affect 9% of adult and yearling reindeer in the Barff herd, and the frequency of occurrence in males and females is similar. Unlike molar damage, mandibular swellings are equally prevalent in young and older age classes (Figure 7.13). Swellings in reindeer are most commonly associated with m1, but a large proportion of the other teeth are involved also (Figure 7.14). Indeed, lesions are associated with the eruption of molars and with the replacement of premolars in animals less than 3 years of age, and with p4 and m1 to m3 in equal frequencies in animals over 3 years of age.

Figure 7.12. Dry mandibles from reindeer of various ages. (1) Normal 19-month-old female with deciduous premolars and two established molars; (2) 13-month-male, with a split m1 and associated large swelling with sinus; (3) 62-month-old female, with missing m1 and large swelling (from Leader-Williams, 1980a). (Photograph by C.J. Gilbert.)

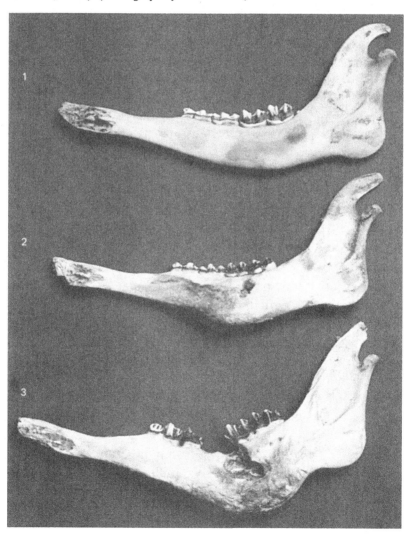

Tooth nomenclature used in Figure 7.14 was after the formula (I0 C1 P3 M3 / i3 c1 p3 m3) x 2 = 34 (Banfield, 1961). Mandibular premolars are termed p2, p3 and p4 and molars m1, m2 and m3. Chi-square tests show there is no difference in the frequency of occurrence of molar damage ($\chi^2 = 0.87$, $df = 1$, $P > 0.10$) or of mandibular swellings ($\chi^2 = 0.67$, $P > 0.10$) in males and females of the Barff herd. Whilst the overall frequency of occurrence of both conditions differs between the three herds (Leader-Williams, 1980a; Chapter 8), there are still no differences

Figure 7.13. Frequency of occurrence of molar damage and mandibular swellings in adult and yearling reindeer with erupting (<3 years of age) and established (>3 years) dentition, summed for all herds. Sample sizes are shown below columns (data from Leader-Williams, 1980a).

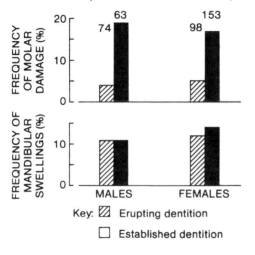

Figure 7.14. Frequency of teeth affected by molar damage and by mandibular swellings in adult and yearling reindeer of both sexes in all herds. p2 to p4 are premolars and m1 to m3 are molars. Sample sizes are shown as appropriate (data from Leader-Williams, 1980a).

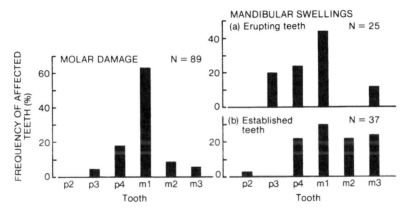

in the pattern of either condition between herds. Data for all herds are thus combined in Figures 7.13 and 7.14. The frequency of occurence of molar damage differs ($\chi^2 = 9.43$, $P < 0.001$) between young (<3 years of age, with erupting teeth) and older reindeer (>3 years with established teeth), but there is no difference ($\chi^2 = 0.15$, $P > 0.10$) in the frequency of occurrence of mandibular swellings in these age groups. The first molar m1 is the tooth most affected in both conditions, but a larger proportion ($\chi^2 = 12.35$, $df = 5$, $P < 0.05$) of the other molariform teeth are associated with mandibular swellings than with molar damage.

The prevalence of mandibular lesions amongst reindeer on South Georgia is unusually high, especially as they are even more common in the Royal Bay and Busen herds (Chapter 8). For example, less than 3.5% of Canadian barren-ground caribou have dental abnormalities, including less serious conditions (Banfield, 1954, 1961; Miller & Tessier, 1971; Miller *et al.*, 1975). Only Alaskan caribou have a prevalence (4–7%) of mandibular lesions approaching that seen on South Georgia (Doerr & Dieterich, 1979). Mandibular lesions are seen infrequently in most populations of ungulates yet occur with a high prevalence in an occasional unusual population (e.g. goats: Rudge, 1970; moose: Peterson, 1977; waterbuck: Foley & Atkinson, 1984; African elephant: Laws & Parker, 1968). These populations have either been introduced and/or resulted from a small founder group (reindeer, goats, moose, waterbuck); or occur at high densities (reindeer, moose, waterbuck and elephants); or occur in areas of unusual mineralisation (reindeer, goats, waterbuck). Genetic traits and inbreeding may contribute to the high prevalence of mandibular lesions (Rudge, 1970; Foley & Atkinson, 1984), but firm evidence is lacking, and would be hard to disentangle from the other influences acting on these unusual populations. Whilst a genetic study would be of interest on South Georgia, inbreeding may well not be of importance because reindeer have high fecundity and low perinatal mortality rates. I now concentrate upon the pathology of mandibular lesions in relation to problems of mineralisation.

7.52   *Pathology*

Soft tissue, bacteriological, radiographic and chemical examination of mandibles was undertaken, first to establish the nature of mandibular swellings, and second to investigate which factors might underlie its high prevalence on South Georgia (Leader-Williams, 1980a). The possible influence of diet was also studied (Leader-Williams, 1982).

Tissue sections were cut and stained from 10% of the lesions after decalcification and bacterial swabs, taken from fresh lesions, were stained and plated out (Leader-Williams, 1980a). Chemical analysis of mandibular bone was undertaken on six diseased and three unaffected mandibles to determine: (a) ash as % of dry fat-free (DFF) bone; (b) ash density (mg mm^{-3}) and percentage of compact bone (VCB %), using volumetric measurements. Radiographic examination was undertaken of ten apparently unaffected mandibles of reindeer of various ages to establish normal patterns of ossification, and of nine diseased mandibles for comparison. A subjective radiographic density score (5 – normal skeletal structure to 1 – severe change) and changes in alveolar bone were noted, and the compact bone percentage (RCB %) measured for each mandible (see Richardson, Richards, Terlecki & Miller, 1979). To clarify the difference between the volumetric and the two-dimensional measurement of compact bone percentage from radiographs, the former is labelled VCB, and the latter RCB. To investigate any relationship between undernutrition and mandibular lesions, plant fragments retained on a 2 mm sieve were taken from the rumens of ten adult female reindeer from each herd in January 1974 and 1975 (Chapter 4). These samples were reduced to ash at 500 °C for 4 h (Allen *et al.*, 1974) as an index of range quality (Klein, 1962).

Actinomycosis ('lumpy jaw') is a bacterial disease of domesticated ungulates (Blood & Henderson, 1979) that occurs also in cervids (e.g. white-tailed deer: Halls, 1984; fallow deer: Chapman & Chapman, 1975). However, lumpy jaw has not been recorded in Eurasian reindeer (Nordkvist, 1971; Nikolaevskii, 1969; Skjenneberg, 1983), nor has it been positively identified in North American caribou (Banfield, 1954; Miller *et al.*, 1975). Mandibular swellings on South Georgia contain dense connective tissue and abscesses of hard, yellow pus. Decalcified sections contain inflammatory cells in which eosinophilic 'sulphur' granules are present and the organisms stain Gram-positive (Figure 7.15). Transverse radiographs of these lesions show evidence of osteoporosis and/or periodontitis (Figure 7.16). Attempts to culture the Gram-positive organisms from bacterial swabs were unsuccessful, probably due to the long period of frozen storage. Present evidence is therefore insufficient to state categorically that mandibular swellings occur as a result of infection by *Actinomyces* sp. and the unambiguous term 'mandibular swelling' is used in place of 'lumpy jaw'.

Radiographic and chemical examinations both show a severe osteoporosis of reindeer mandibles. All mandibles undergo a gradual loss of radiographic density throughout life (Figure 7.17), which reflects changes in mineralisation, matrix and water content. Furthermore, their compact bone % (Table 7.5) falls below the minimum of 40% found in normal sheep mandibles (Richardson *et al.*, 1979). Affected mandibles have even

Figure 7.15. Decalcified tissue sections taken from a mandibular swelling, showing as follows : (A) Encapsulated lesion within mandibular bone, with an associated inflammatory reaction; also rarefying osteitis is evident around the lesion. Stained with haematoxylin and eosin. Scale bar represents 1mm. Arrow indicates position of B and C. (B) Eosinophilic 'sulphur' granules (arrowed) within a granulomatous reaction. Scale bar represents 25μm. (C) Colony of Gram-positive bacteria in lesion. Scale bar represents 10μm (from Leader-Williams, 1980a). (Photograph by C.J. Gilbert.)

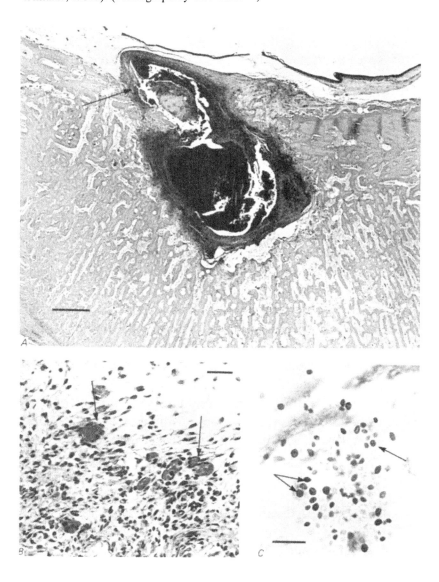

Figure 7.16. Radiograph of transverse sections (2 cm thick) of mandible in the region of premolars p3 to p4. (1) Normal 13-month-old male, with an erupted deciduous tooth and well-developed replacement tooth embedded in thick compact bone, for comparison with (2), a 13-month-old male having a swelling associated with p3: this section shows osteoporosis of alveolar region, poorly developed permanent tooth, and thin compact bone. (3) Normal 63-month-old female with erupted permanent tooth and thick compact bone, for comparison with (4), a 50-month-old female having a swelling associated with a missing p4: this section shows very severe osteoporosis (from Leader-Williams, 1980a). (Photograph by C.J.Gilbert.)

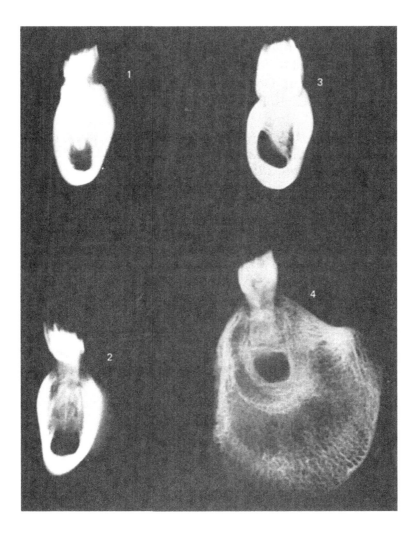

lower radiographic densities and the majority show evidence of osteitis and/or osteoporosis in association with mandibular swellings, either in the region of erupting teeth in younger animals or of tooth roots in older animals (Figures 7.17, 7.18). Different specimens were analysed chemically and mean ash density is similar in control and affected mandibles. Surprisingly, mean ash % DFF bone is higher in affected animals, but both values lie below the normal range of 66–70% found in the metatarsus bones of Norwegian reindeer (Bjarghov, Jacobsen & Skjenneberg, 1977). Most importantly, however, compact bone measured by volumetric methods is lower in affected reindeer (Table 7.5).

In domesticated sheep, osteoporosis is caused by a general dietary deficiency and undernutrition rather than a specific mineral imbalance (cf. McRoberts, Hill & Dalgarno, 1965). Molariform tooth eruption and replacement on South Georgia shows a delay of *c*. 4 months compared with Northern Hemisphere populations (cf. Miller, 1972; Leader-

Table 7.5. *Chemical analysis and radiographic examination of mandibles from reindeer affected with mandibular swellings, and unaffected controls (from Leader-Williams, 1980a)*

Sample	Ash as % of DFF bone	Ash density (mg mm^{-3})	Volumetric compact bone (VCB %)	Radiographic compact bone (RCB %)
Normal	60.3±0.66	1.83±0.28	52.0±1.15	34.4±2.26
N =	3	3	3	10
Affected	63.3±0.49	2.43±0.19	37.3±3.05	23.8±2.22
N =	6	6	6	6

Figure 7.17. Radiographic density scores of unaffected mandibles, of a single mandible with molar damage and of mandibles affected with swellings (data from Leader-Williams, 1980a).

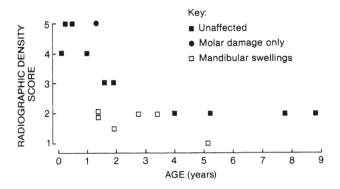

Key:
■ Unaffected
● Molar damage only
□ Mandibular swellings

Williams, 1979a). Radiographs show that compact bone is thin over an area of tooth eruption due to demand by the tooth for metabolites, and that tooth formation is very slow (Figure 7.18). A weak horizontal ramus (Figure 7.16) and loose deciduous teeth predispose to formation of gingival pockets that encourage the entry of infective organisms and the

Figure 7.18. Lateral radiographs of mandibles for which data are shown in Figure 7.17 and Table 7.5. (1) 16-month-old female having a swelling associated with p3, thin compact bone and low radiographic density. (2) 16-month-old female with split m1, but thick compact bone and high radiographic density: apart from the split tooth this mandible appeared normal. (3) 23-month-old female having a swelling associated with p4, no replacement for that deciduous tooth, thin compact bone and low radiographic density (from Leader-Williams, 1980a). (Photograph by C.J. Gilbert.)

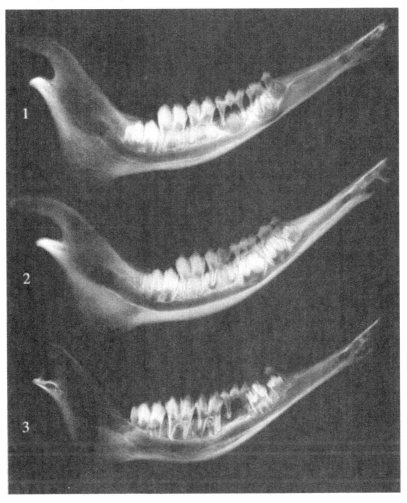

impaction of forage. Similarly, the horizontal ramus becomes less calcified with age (Figure 7.17) and the frequency of tooth loss becomes more common (Figure 7.14), thereby predisposing to infection and to mandibular swellings in reindeer with established dentition. Evidence in support of this chain of events linking a diet generally poor in minerals to osteoporosis and mandibular swellings is provided by analysis of ash content in the rumens of reindeer in each herd on South Georgia. Reindeer with low ash occurred at high densities and had the highest prevalence of mandibular swellings (Figure 7.19).

Percentage results from analysis of mandibular bone were transformed to arcsin values and compared with Student's t test. Mean ash density does not differ ($t_7 = 1.81$, $P > 0.10$) between groups, but values of ash % DFF bone ($t_7 = 3.72$, $P < 0.01$), of VCB % ($t_7 = 3.20$; $P > 0.02$) and of RCB % ($t_{15} = 3.42$, $P < 0.01$) for mandibles affected with swellings differ from those of controls. Spearman's rank correlation coefficient, corrected for ties, shows that radiographic density scores for mandibles unaffected with mandibular swellings decreased with age ($r_s = -0.71$, $N = 11$, $P < 0.05$). Mann-Whitney U test, corrected for ties, shows that reindeer of between 1–6 years of age had lower radiographic density scores ($z = 2.24$, $P < 0.05$) than unaffected reindeer of similar ages. Ash (g % DM) values were ranked and compared by Kruskal-Wallis ANOVA, which shows a difference ($H = 19.16$, $P < 0.001$) between herds.

Figure 7.19. Relationship between prevalence of mandibular swellings, reindeer density and the mean ash content of forage in rumens of ten reindeer from each herd (data from Leader-Williams, 1982).

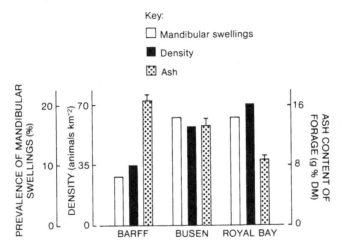

7.53 Effects of the disease

Much is known about the pathology and pathogenesis of diseases in wild ungulates, but few studies demonstrate the ways in which parasites and disease might affect populations (Anderson, 1976; Anderson, 1982). For instance, meningeal worms and virus-induced epizootic haemorrhagic disease have a serious effect upon moose and white-tailed deer (Anderson, 1965; Shope, MacNamara & Mangold, 1960) and rinderpest increases juvenile mortality in African buffalo (Hilborn & Sinclair, 1979). Mandibular swellings on South Georgia provide an opportunity to investigate the impact of a disease upon a wild population of reindeer because lesions can be seen in dead animals, there are no predators to remove evidence, and because information is available on the body condition of affected and unaffected reindeer.

Mandibular swellings are likely to reduce the efficiency of chewing and, thus, the intake of forage (cf. Newton & Jackson, 1983). In turn, this may be expected to cause a reduction in body weights and fat reserves of affected reindeer. Differences between two measures of body condition in affected and unaffected reindeer from all herds are indeed significant for most groups (Figure 7.20). A consistent trend in all herds which showed that females without a calf in summer have a higher

Figure 7.20. Deviations from mean carcass and kidney fat weights of reindeer affected and unaffected by mandibular swellings. * signify a difference of $P < 0.05$ and ** one of $P < 0.01$. Sample size is shown below columns (data from Leader-Williams, 1982).

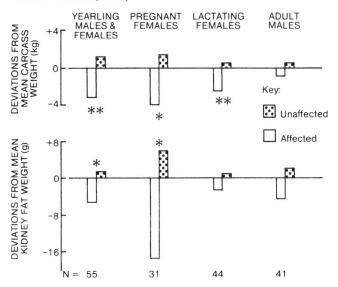

frequency (26%) of mandibular swellings than accompanied females (13%) could not be proven due to the small sample ($N = 7$) of unaccompanied females, but is highly suggestive of a reduction in breeding ability amongst affected females. This conclusion is supported by the reduction in body weight and condition demonstrated amongst most categories of affected reindeer, and by the reduced reproductive performance seen in domesticated sheep with broken mouth (Gunn, 1970).

Mean carcass and kidney fat weights were calculated for seasonal age groups (Leader-Williams, 1982; Leader-Williams & Ricketts, 1982a). Deviations of individual reindeer affected and unaffected by mandibular swellings were ranked, and compared with the Mann-Whitney U test. Because swellings were expected to reduce condition, comparisons were made using one-tailed probability values. Yearling males and females from all herds with swellings ($N = 14$) have lower carcass ($z = 2.68$, $P < 0.005$) and kidney fat ($z = 1.68$, $P < 0.05$) weights than contemporaneous, unaffected reindeer of the same sex. However, adult males with swellings ($N = 12$) have carcass ($z = 0.87$) and kidney fat ($z = 1.23$) weights which do not differ (both $P < 0.10$) from unaffected males. In contrast, both pregnant ($N = 8$) and lactating ($N = 8$) adult females with swellings have lower carcass weights ($z = 2.30$ and 2.35, $P < 0.02$ and 0.01 respectively) than unaffected females; in addition, affected pregnant females have lower ($z = 2.23$, $P < 0.02$) kidney fat weights, but those of lactating females do not differ ($z = 1.27$, $P < 0.10$). Data are insufficient to compare body weights and fat reserves of affected and unaffected females that were unaccompanied by a calf in

Figure 7.21. The numbers of carcasses found with mandibular swellings (hatched column) compared with numbers expected (open column), shown for each herd (data from Leader-Williams, 1982).

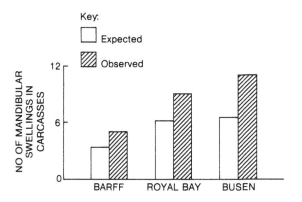

summer months, and to prove ($\chi^2 = 2.34$) that there is a reduction in reproductive performance amongst affected females.

As mandibular swellings debilitate most groups of affected reindeer, they are likely to act as a contributory cause of mortality. If this is so, diseased mandibles would occur more frequently in carcasses than in the shot sample. Indeed, carcasses resulting from natural deaths of adults and yearlings in all herds have more mandibular swellings than is expected (Figure 7.21). This confirms that mandibular swellings are a contributory cause of mortality and demonstrates the possible impact of dental disease and tooth wear upon wild populations of deer (cf. Tyler, 1986).

> The numbers of carcasses resulting from natural deaths expected to have mandibular swellings was derived from frequencies for each herd in the shot sample (Leader-Williams, 1980a). The comparison with numbers of swellings actually observed (Figure 7.21) shows that carcasses have a higher proportion of swellings than is expected ($\chi^2 = 4.38$, $P < 0.05$) for all herds combined.

7.6   Mortality: the effects of ecology and sexual dimorphism

Climatic severity and ecological conditions vary greatly in the wide range of latitudes occupied by reindeer and caribou (White *et al.*, 1981). At the northerly extremity of *Rangifer*'s range at 80°N, Spitzbergen reindeer, tundra reindeer on Novaya Zemlya, Severnaya Zemlya and Novosibirsky islands, and Peary caribou face an extreme and extended winter with little shelter. Sea-ice is the only route of escape from such harsh conditions and, if the sea-ice fails or if winters are unusually severe so that ground-ice or snow accumulation prevents foraging, extensive mortality occurs (Parovshchikov, 1965; Kischinskii, 1971; Miller, Russell & Gunn, 1975; Reimers, 1982; Gates, Adamczewski & Mulders, 1986). Indeed, one High Arctic sub-species of *Rangifer*, the East Greenland caribou, became extinct early this century because of such a climatic shift (Vibe, 1967). However, an advantage of the High Arctic is that predators are either totally lacking, as on Svalbard, or are present only irregularly or at low densities, as in the Canadian Arctic Archipelago, in Greenland and on Russian islands (Kischinskii, 1971; Gossow, 1974; Miller *et al.*, 1977).

In continental areas inhabited by tundra and forest reindeer and by barren-ground, Alaskan and woodland caribou, climate is still an important factor in their ecology. However, in these areas reindeer and caribou have co-evolved with mammalian predators. The most notable predators of wild *Rangifer* are the various species of timber and tundra wolves (Murie, 1944; Skoog, 1968; Kelsall, 1968; Zhigunov, 1969; Parker, 1972, 1973; Miller & Broughton, 1974; Bergerud, 1974c, 1980a, 1983; Klein & Kuzyakin, 1982; Gasaway *et al.*, 1983; Gauthier & Theberge, 1986). In addition, lynxes, or combination of lynxes, wolverines and bears can be locally important as predators (Kelsall, 1968; Bergerud, 1971, 1980a; Semenov-Tian-Shanskii, 1975). These various predators have been present until at least historical times, but have been eliminated from many continental areas by man because they conflict with his interests (Bergerud, 1980a; Nordkvist, 1980; Skogland, 1985). Instead, the spread of firearms and efficient methods of hunting have become important substitutes for natural predation in many areas (Bergerud, 1974c, 1983; Miller, 1983).

Accordingly the patterns and causes of mortality amongst reindeer and caribou might be expected to vary considerably between sub-species and locality. At one extreme most deaths occur as a result of winter starvation and food limitation, and at the other extreme predation, including that by man, is of great importance. In addition, specific diseases and parasites are important causes of mortality in certain areas. For example, protostrongylid nematodes are an important cause of mortality in continental areas, amongst reindeer in Scandinavia and amongst woodland caribou in areas of North America where their ranges overlap with white-tailed deer (Anderson, 1971, 1972; Halvorsen *et al.*, 1980). In contrast, gastrointestinal nematodes are important in the High Arctic, amongst Spitzbergen reindeer (Bye & Halvorsen, 1983). Thus no one cause of mortality is likely to be common to reindeer and caribou in all the situations in which they occur. Most other temperate cervids will face similar variations in causes of mortality in their present ranges, depending on climatic severity and/or the presence or absence of predators, of hunting, of food limitation and of specific diseases (e.g. roe deer: Prior, 1968; Borg, 1970; Bobek, 1977; moose: Peterson, 1955; Peterson, 1977; Gasaway *et al.*, 1983; Messier & Crête, 1984, 1985; black-tailed deer: Wallmo, 1981; white-tailed deer: Halls, 1984).

One factor common to all temperate cervids, however, is the differential mortality which occurs between the sexes (Clutton-Brock *et al.*, 1982). Furthermore, this differential has been found in all situations in which the species has been studied, whether populations are food- or predator-limited, natural or introduced (see references cited previously). It has been suggested that higher mortality rates amongst male juveniles occur as a result of the greater susceptibility of males to food shortage associated with their greater growth and increased nutritional requirements (Clutton-Brock *et al.*, 1982, 1985b). Results from South Georgia suggest this explanation can be extended to include all differential mortality occurring later in life. In the unnatural situation of South Georgia, no differences were detectable between the sexes in the importance of any disease contributing to mortality, including two that were common (internal parasites and mandibular swellings) and studied in some detail. The only factor that correlates with the increased differential between sex ratios with age is dimorphism in body weight (Figure 7.22). Furthermore, the periods at which males and females die (Figure 7.5) coincide with the seasons of weight loss (Figure 6.11). These in turn are related to the differing costs of reproduction in each sex, as measured by the larger proportional loss in fat and body weight incurred by males (Chapter 6). This supports the view that, at every moment in the game of life, the male is playing for the higher stakes (Williams, 1975).

7.7   Summary

1. There is a mortality differential between the sexes and males suffer greater mortality than females.
2. Reindeer calves die either perinatally or during their first winter. Rates of calf mortality are low compared with populations where predators are present.

Figure 7.22. Differences in the sex ratio and in body weight dimorphism amongst temperate cervids. (A) shows differences in sex ratios between birth and first year of life and adult weight dimorphism in various species (based on Clutton-Brock *et al.*, 1985b with additional data from Verme, 1983). 1, roe deer; 2, wapiti; 3, red deer; 4, black-tailed deer; 5, white-tailed deer; 6, reindeer. (B) shows differences between sex ratios at birth and weight dimorphism throughout life in reindeer (based on data from Table 6.3 and Figure 6.9).

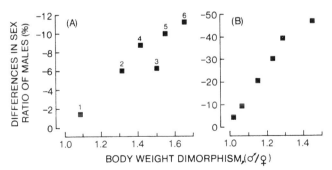

3. Mortality in adults and yearlings follows the pattern typical for mammals. Female reindeer mostly die in late winter, whereas most males die in early winter after the rut, at periods of high reproductive costs to each sex

4. Falls over cliffs are an important cause of death amongst all reindeer, and especially in calves. Disease is mostly uncommon.

5. The internal parasite fauna is impoverished on South Georgia, and the burden insufficient to be a primary cause of mortality.

6. There is an unusually high prevalence of mandibular swellings amongst adult and yearling reindeer, but there is no difference in prevalence between the sexes. Body condition in affected reindeer is poor and the disease is an important contributory cause of mortality.

7. Sexual dimorphism in body weight and cost of reproduction, rather than disease, is the most likely cause of the sex differential in mortality.

Part III

Ecology of an introduction

This section focuses on comparative aspects of the irruption of reindeer. Irruptions occur amongst both established and introduced, and in continental and island, populations of *Rangifer*. A wide range of irruptive responses has been shown by different populations of reindeer and caribou, some exposed to predation and others not. Even where predators are lacking on islands, very different responses have resulted after introduction of reindeer. On certain Arctic islands, introduced reindeer increased rapidly in numbers, but then showed dramatic population crashes. On other islands (including South Georgia), however, the course of the irruptions has been less extreme. These differences provide an opportunity for a comparative study of factors that might determine the course and outcome of irruptions (Chapter 8). On islands where predators are lacking, populations of introduced reindeer are limited by the food supply. As a consequence reindeer have an impact on insular floras. In the case of South Georgia, the vegetation is not adapted to mammalian herbivory, and the extent to which overgrazed vegetation can recover was studied experimentally (Chapter 9). Besides the introduction of reindeer to South Georgia, various other species of ungulates, as well as rodents, lagomorphs and carnivores, have been introduced to far southern islands. These species too have had a considerable impact on native biotas and comparisons are made between the effects of the various mammalian introductions on southern islands (Chapter 10).

8

Irruptions of reindeer

The causes which check the tendency of each species to increase are most obscure.
Charles Darwin (1859)

8.1 Introduction

A truly natural population may be defined as one in which predator, prey and habitat are uninfluenced by human activity (Peek, 1980). Natural populations of ungulates are now virtually impossible to find because man's activity has become so pervasive. As a result of this influence populations of ungulates may show dramatic changes in number. These can result from the elimination of natural predators, from overhunting or from its cessation, from the introduction or the elimination of disease, from range reduction or expansion, or from introductions made to new areas (Rasmussen, 1941; Caughley, 1970; Jordan *et al.*, 1971; Sinclair, 1979; McCullough, 1979; Owen-Smith, 1981; Clutton-Brock *et al.*, 1982; Pellew, 1983).

Reindeer and caribou are a particularly appropriate ungulate with which to examine population responses under different conditions. The genus *Rangifer* is distributed over wide areas of the Arctic where, in different sub-species and populations, it is subjected to various natural and man-induced influences (Chapter 1). On High Arctic islands, natural predators are absent and populations are usually limited by forage availability and climatic fluctuations (Vibe, 1967; White *et al.*, 1981; Thing, 1984; Tyler, 1987a). In continental areas, different populations are subjected to levels of natural predation and hunting mortality that vary from nil to excessive (Bergerud, 1980a, 1983; Syroechkovskii, 1984; Gauthier & Theberge, 1986). Range reduction, logging and forest fires are further variables (Miller, 1976; Klein, 1982; Skrobov, 1984; Darby & Duquette, 1986). As a result of these various influences, irruptions and declines have frequently been documented amongst natural populations of *Rangifer* (e.g. Figures 1.8, 1.9).

Reindeer and caribou have also been introduced to islands and continental areas under a wide range of ecological conditions (Chapter 2). Introductions generally create a situation of great instability, and ungulates normally respond with a single irruptive oscillation (Caughley, 1970). Once they have entered an irruptive oscillation, stable limit cycles to extinction are exceptionally rare amongst other ungulates (Caughley, 1970, 1976a,b). In this context, population responses amongst introduced reindeer on islands are of particular interest, for several have crashed (Figure 2.6), the best documented having been on St Matthew Island (Klein, 1968).

In this chapter, I examine the responses shown by both introduced and natural populations of *Rangifer*. In particular, I concentrate upon factors that might have determined the course of irruptions of reindeer on islands. First I review the fates of the introductions of *Rangifer* to various islands (8.2). Then I discuss factors that might have influenced the rates of increase and the attainment of peak numbers (8.3), and the decline (8.4), of reindeer during irruptions on islands. Throughout these sections I use the opportunities presented for comparisons between the different herds on South Georgia, each at a different stage of their irruption (Chapter 3), and reindeer on St Matthew Island. Finally, I compare factors limiting irruptions in natural and introduced populations (8.5).

8.2   Introduced reindeer on islands

Reindeer (and, less frequently, caribou) have been introduced to a number of islands throughout the world (Table 2.1, Figure 2.4). The islands vary greatly in size, latitude and ecological complexity. They range from large islands like Baffin, Iceland or Southampton to small islands like the Aleutians or Pribilofs. Additionally, large islands such as Greenland or South Georgia may be divided into 'small islands' by glaciers. The Northern Hemisphere islands also range in latitude from 40° to 77°N, and in habitat type from boreal forest to High Arctic tundra. Thus all the introductions (with the obvious exception of those to southern islands) encompassed latitudes and habitat types normally occupied by *Rangifer* (cf. Figure 1.2 with Figure 2.4). In some cases, semi-domestic reindeer (or caribou) were reintroduced into areas or islands (e.g. Baffin, Greenland, Newfoundland, Nunivak, Scotland and Southampton) that had been occupied in recent or historic times by wild reindeer or caribou. In many cases, however, the introductions were made to islands (e.g. Aleutians, Iceland, Kerguelen, Pribilofs, South Georgia) that were not known to have been occupied previously by *Rangifer*, at least in historic times. In this section I first review possible reasons for success and failure of these introductions, and second examine their trends of numerical change.

8.21   Failure and success of introductions

A total of 31 introductions of reindeer (or caribou) to different islands have been traced, including six made to different areas of Greenland and two to Iles Kerguelen that are not listed separately in Table 2.1. The fate of most of the introductions is known either from anecdotal accounts, or, less often, from more detailed studies. An assessment of the fates of the introductions (Figure 8.1) shows that several (16%) failed at source, whereas the remainder entered an irruptive oscillation (cf. Figure 2.1). Introductions appear to have failed at source for completely different reasons. The mistaken identification by the herders of mosses for lichens, and the consequent lack of suitable

Figure 8.1. Outcomes of introductions of reindeer or caribou to different islands or areas of islands throughout the world (based on data from Table 2.1).

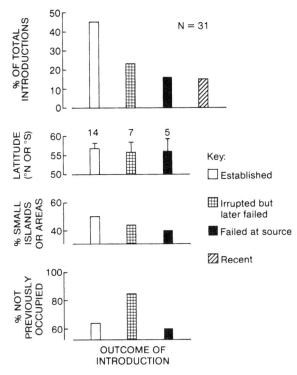

forage, was blamed as the cause of failure of the introduction to Baffin Island in 1921 (Grenfell, 1934). On Great Cloche Island, reindeer were infested by the parasitic nematode *Pneumostrongylus tenuis* and died from the resulting neurological symptoms and paralysis (Anderson, 1971). On South Georgia one introduction was wiped out by a snowslide (Olstad, 1930). Unfortunately, the causes of failure of the introductions to Iceland in 1771 and to Isla Navarino in 1944 are not documented.

Of the introductions that irrupted, five were made between 1965 and 1973 to different areas of east and west Greenland from the original herd introduced to the Itinnera area in 1952 (Strandgaard *et al.*, 1983) and it is probably too soon to assess their eventual outcomes. Many (45%) of the other introductions have resulted in established populations, but a large proportion (23%) failed after an initial irruption. Neither the latitude nor the size (and presumably related factors like temperature or number of plant species), nor previous occupancy by *Rangifer* has any obvious relationship to the subsequent success or failure of the introductions (Figure 8.1). Therefore I will now concentrate upon more detailed data from islands where these are available, with the aim of explaining the marked differences in the eventual outcomes of the introductions that irrupted.

> Accounts of the outcomes of introductions of reindeer to islands (Table 2.1) were interpreted and classified into four categories (Figure 8.1). Detailed studies could be categorised without difficulty. In anecdotal accounts, phrases like 'were at one time common' and so on, were taken to mean that an irruption had taken place. Such interpretations are not infallible, but unlikely to be far wrong. To test for possible differences in characteristics for islands, it was necessary to combine the small samples from the 'irrupted but later failed' and 'failed at source' categories (Figure 8.1). A Mann-Whitney U test shows that there is no difference in latitude between the islands or areas on which introductions have become established and on which they failed ($U = 57$, $n_1 = 12$, $n_2 = 14$, $P > 0.10$). Chi-square tests show that there is no difference in the proportions of large to small islands ($\chi^2 = 0.18$) or of formerly occupied to unoccupied islands ($\chi^2 = 0.34$) on which introductions succeeded or failed (both $P > 0.10$).

8.22 Numerical change

Introductions that do not fail at source normally adjust to their new environment with a single irruptive oscillation (Figure 2.1). Historical records of reindeer numbers are available for seven islands and for two herds on South Georgia, as shown individually in Figures 2.6 and

3.9. In addition, data from South Georgia, St Paul and St Matthew islands are superimposed in Figure 8.2 to emphasise the differences in the irruption of each herd. The various irruptions appeared to differ in three important respects: (1) in the maximum numbers attained; (2) in the number of years after introduction that these peaks occurred; (3) most especially, in the nature and rate of their declines (Figure 8.2).

In the absence of predators, introduced ungulates initially increase in number in response to the discrepancy between the carrying capacity of their new environment and the numbers actually present (Figure 2.1). The total area accessible to reindeer on each island is shown in Table 8.1, but these data take no account of the area of productive or vegetated range available for use by each herd. If the peak numbers attained are considered in terms of density per total area of range (Table 8.1), the maxima were very high (18–23 km^{-2}) for the Barff, St Matthew and St Paul herds, high (6 km^{-2}) for the Busen and Nunivak herds, and of similar densities (0.3–2 km^{-2}) to those seen amongst continental herds (Figure 3.14) for the Adak, St Lawrence and St George introductions. Scheffer

Figure 8.2. Summary of reindeer numbers on South Georgia, St Paul Island and St Matthew Island, from data in Figures 2.6 and 3.9 (from Leader-Williams, 1980d).

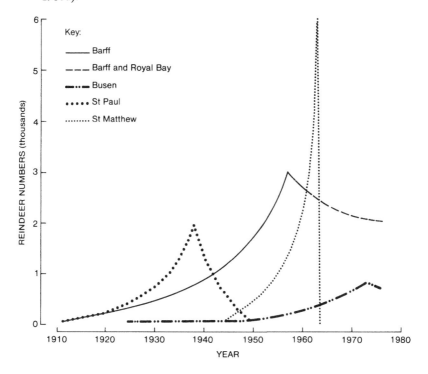

(1951) could advance no reasons for the difference between the neighbouring St Paul and St George islands and no further data, such as productivity or the area grazed by each herd, are available to explain the large range of peak densities attained on the various other Arctic islands. On South Georgia, however, the Busen herd had not negotiated four mountain passes leading into potentially available, but as yet unoccupied range within their overall area (Figure 3.13; see also Figure 9.2). If density is calculated using only occupied and vegetated areas grazed by reindeer (Table 3.3), peak densities were similar in the Barff (100 km^{-2}) and Busen (71 km^{-2}) herds, and close to the 1976 density (85 km^{-2}) seen in the Royal Bay herd, which at that time probably was approaching or had reached maximum numbers attained after emigration to this new area. Thus on South Georgia, where different reindeer areas were likely to have had comparable productivities, the maximum numbers attained by introduced reindeer generally have been proportional to the area available.

The increase in numbers in each introduced herd appears to have been exponential (Figures 2.6, 8.2), though there are relatively few estimates for some herds. The length of the period between introduction and attainment of maximum numbers will have depended on the carrying capacity of each island for reindeer (above), on the initial number introduced (Table 2.1), and on the rate of population increase. An analysis of the rates of increase shows differences between islands (Figure 8.3). On Arctic islands rates of increase varied from 0.13 on Adak to 0.28 on St Matthew, close to the theoretical maximum for reindeer of 0.30 (White *et al.*, 1981). On South Georgia the rate of

Table 8.1. *Peak densities attained by introduced reindeer on different islands*

Island/herd	Total area (km^2)	Peak density (animals km^{-2})
Barff	131	23
Busen	124	6
Adak	750	0.3
Nunivak	3500	6
St Lawrence	4100	0.4
St Matthew	332	18
St Paul	107	19
St George	91	2

increase in both herds has been low (0.10–0.11) but comparable with that seen on St Paul and Adak islands (Figure 8.3).

An exponential curve was fitten by linear regression of $\log_e$ reindeer numbers on year, from the date of introduction to attainment of peak numbers, for eight islands with four or more estimates (Figure 8.3). Correlation coefficients for the regression lines are all significant, and the slope of each line represents the instantaneous rate of increase. A common line was fitted, but accounted for only 39% of the variance ($F_{1,39} = 208.62$). However, inclusion in the regression model of separate intercepts ($F_{7,32} = 33.69$) and slopes ($F_{7,25} = 5.77$, $P < 0.005$) accounted for 92% of the variance. Thus, rates of increase differed between herds. The rate of increase in the Barff herd did not differ from that in the Busen, St Lawrence, St Paul or Adak herds (all $P > 0.05$), but was lower than that in the Nunivak ($t = 3.90$, $df = 25$, $P < 0.001$), St George ($t = 2.16$, $P < 0.05$) and St Matthew herds ($t = 4.47$, $P < 0.001$).

After increasing in number to the full carrying capacity of the range, introduced ungulates usually follow a sequence of a levelling-off in numbers, a decline, and then a phase of relative stability in which

Figure 8.3. Rates of population increase of reindeer introduced to islands. An exponential curve was fitted for each herd, and its slope (m) represents the instantaneous rate of increase between introduction and attainment of peak numbers. Data from St George Island are omitted for clarity, but this herd increased with $m = 0.23$ between 1911 and 1922 (based on data from Figures 2.6 and 3.9).

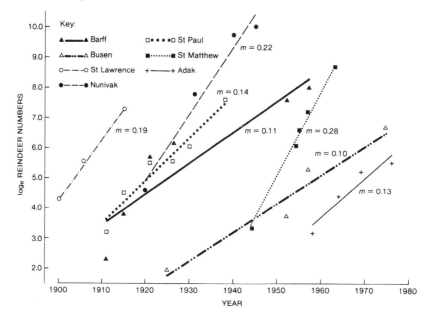

population density remains lower than peak density (Figure 2.1). Introduced reindeer populations on South Georgia, St Lawrence and St
George islands appear not to have deviated from this pattern (Figures
2.6, 8.2). Reindeer on St George Island decreased at a rate of −0.06 per
annum from their peak and now occur at even lower densities. The rate
of decrease was −0.05 in the Barff herd, but emigration of reindeer to the
Royal Bay area has complicated the picture. On St Paul and Nunivak
islands the declines were more exaggerated with rates of decrease of
−0.37 in the former and loss of *c*. 75% of the population in a crash on the
latter: reindeer, however, are still present on both islands at lower
densities. Similarly, an introduction of reindeer to Iceland in 1787

Figure 8.4. The course of the irruption of reindeer introduced to Iceland in 1787.
Maps show the distribution of reindeer during their initial irruption, whilst maps
and counts show the recovery of reindeer during a later irruption (based on data
from Thórisson, 1983).

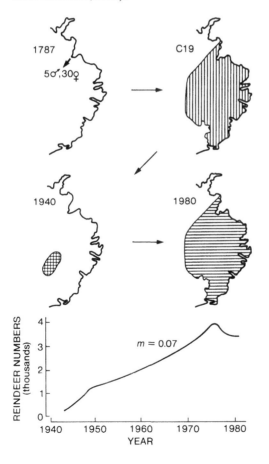

underwent its initial irruption in the 19th century, declined greatly in distribution (and, by inference, numbers), but has now begun to increase again in a phase of greater stability (Figure 8.4). The most exaggerated decline of all was the crash on St Matthew Island, in which over 99% of the population was lost in one year (Figure 8.2).

The range of responses shown by introduced reindeer on islands is also of interest in relation to arguments over limiting factors in continental populations. As levels of natural predation and/or hunting mortality increase in continental populations of caribou and reindeer (shown in Figure 8.5 by moving across the central nine cells from top left to bottom right), populations move from high mean rates of increase to rates of decrease. Thus it has frequently been argued that predation alone provides the limiting factor for continental *Rangifer* populations and that

Figure 8.5. The mean ± SE rates of increase or decrease amongst island herds of introduced reindeer, compared with those seen amongst continental herds of *Rangifer*. The nine central cells show data from 46 continental herds in North America and Europe subjected to three levels of natural predation and/or hunting mortality (based on Bergerud, 1980a, with additional or corrected data from Figures 1.8, 2.5; Krebs, 1961; Semenov-Tian-Shanskii, 1975; Klein & Kuzyakin, 1982; Danilov & Markovsky, 1983; Skogland, 1985; Williams & Heard, 1986). The left outer cell shows rates of increase of eight introduced island herds (Figure 8.3), whilst the right outer cell shows rates of decrease from five of the same islands during the decline phase (Figures 2.6, 3.9).

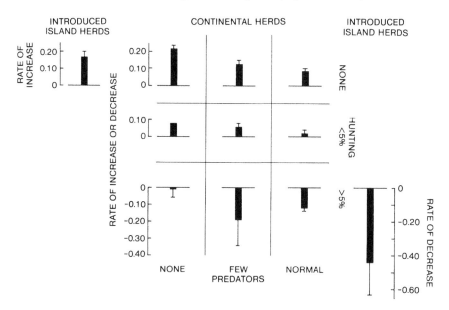

food limitation is of no importance (Bergerud, 1974c, 1980a, 1983). In contrast, some introduced island herds are harvested, but this has not been universal, and natural predators are absent. Hence predation cannot be invoked as the limiting factor on islands, where equally marked and highly variable changes occur in rates of increase (Figure 8.5). In this situation, the upswing in an irruption appears to be initiated in response to an increased availability of food and is terminated by overgrazing (Caughley, 1970, 1976a,b). At high latitudes there are two aspects of forage availability that must be considered, namely the carrying capacity of the winter range and the productivity and quality of the summer range (Klein, 1970a; White *et al.*, 1981). These two aspects of range quality will be used together with the three aspects of numerical change outlined above, to identify factors of importance in the irruptions of reindeer on islands.

> Rates of increase and decrease in continental herds subjected to different levels of predation (Figure 8.5) were mostly taken from Bergerud (1980a), but with some modification and with additional data from Eurasian herds. Kruskal-Wallis ANOVA shows that rates of increase or decrease differ amongst the 46 herds when divided into nine cells according to levels of predation ($H = 29.37$, $df = 8$, $P < 0.001$). Mann-Whitney U test shows there is no difference in rates of increase between island populations and the adjacent cell ($U = 9$, $n_1 = 4$, $n_2 = 8$, $P > 0.05$).

8.3   Increase and attainment of peak numbers

The age of first breeding and rates of fecundity are of importance in determining intrinsic rates of increase of a population, and ungulates often vary their reproductive rates. Thus age-specific fecundity rates of Himalayan thar differed at each stage of their irruption in New Zealand (Caughley, 1970). Amongst cervids, moose, roe, white- and black-tailed deer respond to optimal or unfavourable conditions by altering their ovulation and multiple birth rates (Chapter 4; Klein, 1970b, 1981). However, changes in the fecundity of reindeer and other species which normally bear single young usually arise from the occurrence of calf pregnancies under optimal conditions or from reduced fertility in yearling and young adult females under unfavourable conditions (Chapter 4; White *et al.*, 1981). Therefore *Rangifer* has less phenotypic plasticity to alter its reproductive strategy than species that have multiple ovulations and births. In this section I examine factors that might have influenced rates of fecundity and limited population growth on different islands.

8.31   Fecundity

No data are available on the fecundity rates or age of first breeding of introduced reindeer on islands other than South Georgia, though ovarian examination suggested that calves bred on St Matthew Island (Klein, 1968). Data from South Georgia are available only for 1973–6, but they actually provide a wide range of information due to the early stage of the irruption in the Royal Bay herd. In spite of this, there are no detectable differences in age-specific or overall fecundity rates between the herds, and pregnancies do not occur in calves. Fecundity rates are high (>90%) in reindeer over 18 months old in all herds (Table 4.3). Examination of ovaries confirms that there are no structures suggestive of calf pregnancies in the Royal Bay herd and the regression of luteal scars on age does not differ between Barff and Royal Bay herds (Leader-Williams & Rosser, 1983). These observations, coupled with known instantaneous rates of increase of 10–11% (Figure 8.3), suggest that females have never attained puberty as calves on South Georgia, in contrast to St Matthew where calf pregnancies are likely to have permitted a high rate of increase. The lack of any difference in fecundity on South Georgia is perhaps surprising, because fecundity rates and body-size of Norwegian reindeer varies in herds of different densities and on different qualities of range (Reimers, 1972; Reimers *et al.*, 1983; Skogland, 1983, 1985).

> Chi-square tests show that there are no differences ($\chi^2 = 0.40$, $df = 2$, $P > 0.10$) in pregnancy rates between herds (Table 4.3). No luteal structures were found in the ovaries of yearling reindeer ($N = 11$) in the Royal Bay herd, confirming a lack of pregnancies in calves. The regression of luteal scars on age for the Royal Bay herd is significant ($r = 0.88$, $P < 0.001$) and of the form $y = -0.72 + 1.01x$. Neither the slope ($F_{1,192} = 1.49$, $P > 0.10$) nor intercept ($F_{1,193} = 3.81$, $P > 0.10$) differ from the same regression for the Barff herd (Figure 4.11).

8.32   Growth

Body weight and condition has a marked effect on fecundity rates (Chapter 6), and increasing population density is usually associated with a depression of growth rate and body size (Scheffer 1955; Skogland, 1983). Thus changes in body weight and condition most probably underlie any differences seen in fecundity during an irruption (Caughley, 1970). On South Georgia, reindeer from the Busen herd are consistently larger than those in both other herds (Leader-Williams & Ricketts, 1982b). However, their ancestors derive from different genetic stock to the Barff introduction (Table 3.3), and it is not possible to separate phenotypic from genotypic differences in body size. Fortunately, a wide range of data are again available from comparisons of the genotypically identical Royal Bay and Barff herds for comparison with data from St Matthew Island (Klein, 1968).

Comparative data from the two herds were collected in December and January, at times in the annual cycle when both males and females have low body weights and when antlers are in velvet (Chapter 6). Males from the Royal Bay herd show a significant growth advantage over their counterparts in the Barff herd: they have longer crown–tail and jaw lengths and heavier body and leg weights (Figure 8.6). However, little growth advantage accrues to females, and only their jaw lengths are greater in the Royal Bay herd. In addition, velvet antler weights of males

Figure 8.6. Comparison of lengths and weights of reindeer at early and late stages of their irruption on South Georgia. Asymptotic skeletal measurements and body and organ weights are shown for males and females from the Royal Bay and Barff herds during early summer. Significant differences between herds are shown with * = $P < 0.05$, ** = $P < 0.01$ and *** = $P < 0.001$ (data from Leader-Williams & Ricketts, 1982b).

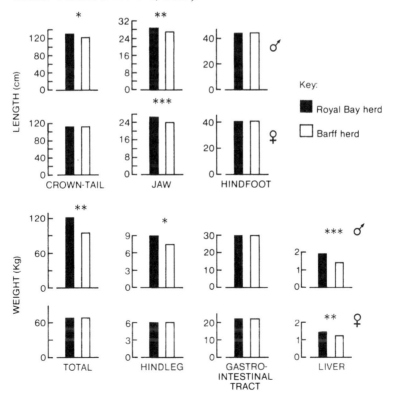

Figure 8.7. Reindeer on St Matthew Island in July 1957. Males, females and calves all have heavy body weights and well-developed antlers. A deep mat of lichens is visible in the foreground. (Photograph by D. R. Klein.)

are greater in the Royal Bay herd, but antlers of females are the same weight in both herds. These inter-herd comparisons amplify and confirm an intra-herd comparison based on a small data set from 1928, when the Barff herd was expanding (Olstad, 1930). Little advantage in growth accrued to females even in 1928, but there was a large difference in weights and length of males in 1928 compared with 1975 (Leader-Williams, 1980d). Direct comparisons can be made between the size of reindeer on South Georgia and on St Matthew Island (Klein, 1968). Males were much longer and heavier on St Matthew in 1957, when the herd was expanding, than in 1963, immediately before the crash. However, and in contrast to South Georgia, St Matthew females also had heavier total weights and longer total and hindfoot lengths in 1957 (Figure 8.7). Even though no firm data are available on the age of first breeding on St Matthew Island, the supposition that calves became pregnant when the herd was expanding is supported by comparisons of growth on the two islands. On South Georgia, relatively small differences in female growth at early and late stages of their irruption have not resulted in calf pregnancies. However, the large differences in female growth on St Matthew Island most probably would have allowed calves to reach the size threshold for puberty or conception (Chapter 6), thereby explaining the high rate of increase of this population (Figure 8.3).

> Reindeer lengths and weights were measured in the Royal Bay and Barff herds in December and January 1973–6, as described in Chapter 6. Data were analysed using a growth curve fitted by weighted least squares (Leader-Williams & Ricketts, 1982b). Analysis of variance shows there are the following differences in asymptotic lengths or weights for males in the two herds: crown–tail length: $F_{1,36} = 4.82$, $P < 0.05$; jaw length: $F_{1,37} = 8.89$, $P < 0.005$; total weight: $F_{1,8} = 18.79$, $P < 0.005$; hindleg weight: $F_{1,8} = 9.63$, $P < 0.025$; liver weight: $F_{1,8} = 29.05$, $P < 0.001$). There are the following differences for females: jaw length: $F_{1,80} = 12.16$, $P < 0.001$; liver weight: $F_{1,13} = 12.10$, $P < 0.005$. Klein (1968) analysed data from St Matthew Island somewhat differently, but showed differences for females in their total weights ($P < 0.025$), total lengths ($P < 0.001$) and hindfoot lengths ($P < 0.025$).

Reindeer on St Matthew Island were shown to have decreased in size as herd density increased (Klein, 1968). However, the direction of the difference in size of reindeer in the Royal Bay and Barff herds might at first appear surprising, for the larger animals occur in the higher density herd, contrary to general predictions (Scheffer, 1955). In territorial species of cervid, such as roe deer, variations in body size are directly related to population density (Klein & Strandgaard, 1972; Bobek, 1977). In contrast, non-territorial species such as *Rangifer* are primarily affected by the availability of their preferred forage species, and not necessarily by density itself (Klein, 1970a, 1981; Ricklefs, 1973).

Figure 8.8. Comparison of the summer diet of reindeer at different stages of their irruption on South Georgia. Mean proportions of each forage species present in rumen contents during January are shown for each herd, with sample sizes above columns. The proportions of preferred species in the diet decreases from left to right, as shown by selectivity scores (based on data from Leader-Williams *et al.*, 1981).

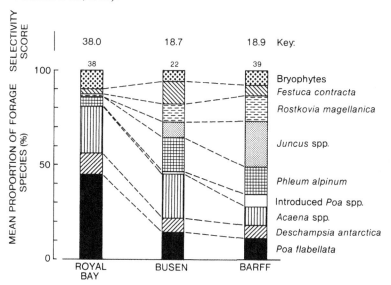

Figure 8.9. Comparison of ruminal nitrogen during early summer in reindeer on South Georgia and on Arctic islands during different stages of their irruptions. Data are shown from January for three southern herds and from July for Arctic island herds at early and immediately pre-crash stages of their irruptions (data from Klein, 1968).

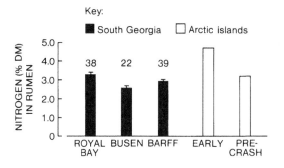

8.33   Summer forage

Reindeer have occupied each area of South Georgia for different lengths of time (Table 3.3). The Royal Bay area has been most recently occupied and the Barff Peninsula has been grazed for longest. Thus reindeer have had the greatest impact on preferred forage species on the Barff (Chapter 9). This is shown by comparison of rumen contents of reindeer in different herds in early summer (Figure 8.8). There are clear differences in the proportions of the various species in the diet of each herd. However, these differences might also be related to variations in total areas of each community available to the three herds, for which there are no data for the Busen and Royal Bay areas. To circumvent this problem, I have instead calculated a selectivity score for each herd, thereby demonstrating that the summer diet of Royal Bay reindeer contains a high proportion of preferred species (Figure 8.8). I have already established that reindeer on South Georgia select forage species during summer largely on the basis of their nitrogen contents (Chapter 5). As might be expected, therefore, ruminal nitrogen contents differ and are highest in the Royal Bay herd (Figure 8.9). Thus liver weights, which are particularly sensitive to changes in diet quality (Chapter 6), are also heavy in males and females from this herd (Figure 8.6).

Comparisons of the quality of vegetation available to introduced reindeer on different islands are limited. On South Georgia, reindeer encountered a species-poor vascular flora that was not adapted to grazing and was dominated by species of low nutritive quality but high productivity (Lewis Smith, 1984). In contrast, reindeer on St Matthew Island encountered a richer flora comprising over 100 species (Klein, 1959). Several species had similar nitrogen contents to those available on South Georgia, but the relative abundance of sedges, *Carex* spp., and other high quality forage was high, and grazing stimulated further production (Klein, 1968). As a result, the nitrogen content of forage in the rumens of reindeer was generally higher on Arctic islands than on South Georgia (Figure 8.9). This fundamental difference most probably underlies the contrasts in growth shown by females on South Georgia and St Matthew Island, and hence in their different rates of population increase. It is a clear example of the multiplier effect, in which small differences in range quality result in large differences in productivity (White, 1983).

Rumen samples collected in January 1974–6 in the Barff, Busen and Royal Bay herds were analysed as described in Chapter 5. Samples from the Royal Bay herd contain only trace qualities of *Juncus* spp. and introduced *Poa* spp., but comprise more *Poa flabellata* ($t_{75} = 9.00$, $P <$

0.001), *Deschampsia antarctica* ($t_{75} = 2.52$, $P < 0.02$) and *Acaena* spp. ($t_{75} = 8.62$, $P < 0.001$), and less *Phleum alpinum* ($t_{75} = 3.14$, $P < 0.01$) and *Rostkovia magellanica* ($t_{75} = 10.93$, $P < 0.001$) than Barff samples. Samples from the Busen herd comprise more *Acaena* spp. ($t_{59} = 7.30$, $P < 0.001$) and *Festuca contracta* ($t_{59} = 5.64$, $P < 0.001$), and less *Juncus* spp. ($t_{59} = 5.94$, $P < 0.001$) and introduced *Poa* spp. ($t_{59} = 3.02$, $P < 0.01$) than Barff samples. A selectivity score was calculated for each herd as $\Sigma(p_i \times S_i)$, with p_i = mean proportion of the *i*th species (Figure 8.8), and S_i = index of selectivity of that species (Table 5.4). The nitrogen content of rumen samples was analysed as described in Chapter 5. Kruskall-Wallis ANOVA shows that ruminal nitrogen concentrations differ between herds ($H = 27.34$, $df = 2$, $P < 0.001$) during early summer months.

8.34   Density and carrying capacity

Some island populations of introduced reindeer have attained very high peak densities (Table 8.1). It has been argued that natural predation and hunting limits continental populations of *Rangifer* to densities (≤ 1 km^{-2}) that are well below the carrying capacity of the range, thereby giving rise to the suggestion that densities up to twenty

Figure 8.10. Regression of *Rangifer* biomass on standing crop of vegetation in various Arctic sites and on South Georgia (based on Skogland, 1980, with additional data for South Georgia and a common regression for northern and southern populations).

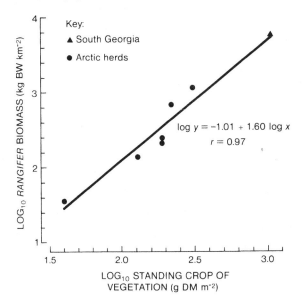

times higher have only been achieved on islands due to the absense of predation (Bergerud, 1974c, 1980a). But are densities achieved by introduced reindeer not related at all to forage availability, or to the less easily defined concept of carrying capacity (cf. Caughley, 1979)?

It is unfortunate that no comparative data on densities of reindeer per km^2 of vegetated range or on standing crop of vegetation are available for other introduced populations. However, the biomass (derived from stocking rates and body weights) of natural populations of *Rangifer* in different areas of the Arctic correlates with the standing crop of vegetation available in summer (Skogland, 1980). Subantarctic tundra is highly productive: the ungrazed vegetation encountered by reindeer on South Georgia comprises the exceptionally productive tussock grassland community (Table 5.1) and other communities of a similar level of productivity to those found in Arctic tundra (French & Smith, 1985). Comparison of peak reindeer biomass found in the Royal Bay herd with the overall standing crop of vegetation likely to have been encountered fits the correlation of Arctic areas (Figure 8.10). This shows clearly that the high densities attained by some introduced reindeer are not simply a default arising from lack of predation, but instead are related closely to availability of forage.

> Data on the biomass of *Rangifer* and standing crop of vegetation in different areas of the Arctic were extracted from Skogland (1980). Comparable data from South Georgia comprised: (1) mean body weight = 64 kg (derived from 100 reindeer shot in Royal Bay, and corrected according to the observed sex structure); (2) density in occupied and grazed range = 85 km^{-2} (Table 3.3); (3) overall standing crop of vegetation in ungrazed areas = 1100 g DM^{-2} (derived from standing crop estimates in Table 5.1 and from the relative area of tussock grassland to other communities in Table 5.2). The correlation of $\log_{10}$ *Rangifer* biomass on $\log_{10}$ standing crop of vegetation was significant ($P < 0.001$) for Arctic areas and South Georgia (Figure 8.10).

8.4   Decline from peak numbers

As an introduced island population of ungulates approaches peak density, then declines and finally stabilises in number (Figure 2.1), the main factor influencing the rate of population increase is variation in rates of mortality (Caughley, 1970). The onset of a population decline is initiated by overgrazing and the transition to a phase of relative stability will occur only if a balance between continued availability of forage on an overgrazed range at critical seasons and the lower density of animals on that range is soon reached (Caughley, 1970, 1976a,b). In this section, I first examine differences in the winter diets of introduced reindeer and then establish whether populations may be food limited. I then examine how patterns and rates of mortality vary during the course of an irruption and examine the influence of climate and emigration on declines of introduced reindeer.

8.41   *Winter forage*

Information on the winter diet of introduced reindeer is available for only a few islands. On South Georgia, *Poa flabellata* now forms the almost exclusive component of the winter diet (Chapter 5) and, as it dominates the only community available to reindeer in winter, there seems no obvious reason why *P. flabellata* has not always been the principal winter forage (Olstad, 1930; Bonner, 1958). Tussock grassland has, in common with other native plant communities, responded unfavourably to grazing and been progressively depleted over long periods by reindeer (see Figure 9.4), but no data are available for present standing crop in grazed communities. However, an earlier correlation (Figure 8.10) suggests that high densities of reindeer are still found at later stages of their irruption in the Barff and Busen herds because of the continued high productivity of grazed tussock grassland (see Figure 9.1A) relative to communities found in Arctic tundra.

But what of winter forage on Arctic islands? Due to their heavy use and distribution on exposed areas with little snow, lichens were suggested to have been the principal winter forage on St Paul and St Matthew islands (Scheffer, 1951; Klein, 1968). Thus reindeer attained high densities on these Arctic islands because of a high rate of increase, an

Figure 8.11. Winter diet of introduced reindeer in Iceland in an area that has been recently colonised compared with one that has been occupied for several decades (data from Egilsson, 1983).

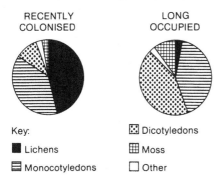

RECENTLY COLONISED   LONG OCCUPIED

Key:

▣ Dicotyledons

■ Lichens   ⊞ Moss

▤ Monocotyledons   ☐ Other

initially abundant supply of lichens (Figure 8.7) and a period when winters were mild. Lichens, however, recover very slowly from grazing and trampling by reindeer (Chapter 9) and densities could not be maintained because no other forage appeared to have been available on these islands: rapid declines then occurred in severe winters (Klein, 1968).

Some Arctic island populations of introduced reindeer have shown slower declines, as opposed to crashes. Furthermore, reindeer in Iceland are now undergoing a second irruption (Figure 8.4). Whilst it is clear that lichens are preferred in newly recolonised areas, Icelandic reindeer do switch to other forage species, notably winter-green vascular plants, when lichens are overgrazed (Figure 8.11). No data are available from other islands, but those from South Georgia and Iceland demonstrate the necessity for introduced reindeer populations on islands to depend upon vascular species in winter if they are to avoid population crashes.

8.42   Condition

Reindeer and caribou live in a highly seasonal environment and have adapted to changes in forage quality and quantity by showing marked cycles in body weight and condition (Chapter 6). However, if a population is indeed limited by the availability of winter forage, it is likely to show even more extreme cycles of body condition than populations limited by other factors, such as predation. There are no data from other islands on changes in body condition of introduced reindeer over an annual cycle. However, comparable studies have been undertaken on South Georgia and on barren-ground caribou in continental Canada (Dauphiné, 1976). In contrast to the reindeer on South Georgia, the caribou ate lichens extensively in winter and were subject to both fly harassment and predation (Parker, 1972; Miller, 1976). Like the Barff herd, however, this herd had decreased in numbers at the time of study.

To facilitate a comparison, data on body weight and fat reserves were extracted from Dauphiné (1976) and reanalysed: direct comparisons are possible for total weight and for femur marrow fat, the last fat reserve to be mobilised (Chapter 6). Although barren-ground caribou have heavier asymptotic body weights than reindeer on South Georgia, the proportional fluctuations in body weight (15% for males and 8% for females) are remarkably similar (Figure 8.12). This provides additional evidence that these fluctuations are sex-specific and closely tied to reproductive activity (Chapter 6). Caribou also have a higher average level of femur marrow fat (64–65% for both sexes) than South Georgia reindeer, but the most striking difference is in their fluctuations (Figure 8.12). The proportional changes in femur marrow fat are only 5% in male and 19% in female caribou, but 45% for both sexes in the Barff herd.

It appears, therefore, that changes in the fat reserves of this population of caribou were insufficient to implicate malnutrition as a cause of mortality (Dauphiné, 1976). In spite of fly harassment limiting their opportunities for fat deposition during summer, caribou still had higher fat reserves than reindeer on South Georgia. The caribou population was at that time limited by natural predation and hunting, and was unlikely to have overgrazed either its summer or winter ranges (Parker, 1972;

Figure 8.12. Proportional fluctuations in body weight and femur marrow fat in male and female *Rangifer* in Canada and on South Georgia. For each sex the upper pair of graphs represent data from South Georgia (Barff herd) and the lower pair represent data from barren-ground caribou (Dauphiné, 1976). The time scale represents months after the mean date of birth of calves (from Leader-Williams & Ricketts, 1982b).

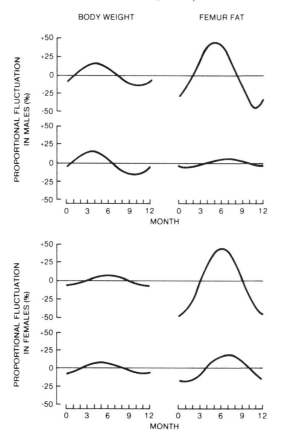

Miller, 1976). By contrast, the unusually large fluctuations in femur marrow fat, the severe depletion of fat reserves and utilisation of body protein by reindeer on South Georgia (Chapter 6) indicate that loss of condition is an important cause of mortality and that this introduced population is indeed food limited.

> Data on total body weights and femur marrow fat % of barren-ground caribou were extracted from Dauphiné (1976) and analysed using a sine wave formula, as described in Chapter 6 and in Leader-Williams & Ricketts (1982b): results are shown only for the proportional fluctuation (*C*) about the growth curve (Figure 8.12).

8.43 Mortality

How then might decreased winter range carrying capacity have acted on patterns and rates of mortality in a food limited population? On South Georgia all reindeer herds have either probably reached peak numbers (Royal Bay) or declined from peak numbers (Busen and Barff), and sex differences in the timing of mortality coincide with the periods when fat reserves and body weights are at their lowest (Chapter 7). When ungulate populations are close to carrying capacity, as are the reindeer on South Georgia, density-dependent factors may be of importance (Sinclair, 1979; Houston, 1982; Clutton-Brock *et al.*, 1985a; Skogland, 1985). Of the measures of mortality made in this study, no differences are evident in the expectation of life of females in each herd, but significant differences occur in rates of perinatal mortality, in the frequency of mandibular swellings and in adult sex ratios (Figure 8.13). Thus as reindeer have moved through the stages of their irruption from peak to lower densities, perinatal mortality (Table 7.2) has increased and adult sex ratios (Table 8.2) have become more biased, indicating a higher mortality of males. Reduced viability in males relative to females when subjected to food shortage or high population density is common amongst ungulates (Robinette, Gashwiler, Low & Jones, 1957; Flook, 1970; Clutton-Brock *et al.*, 1982, 1985a,b; Houston, 1982; Skogland, 1985). Furthermore, the study provides an unusual example from an ungulate of a disease condition that acts as a contributory cause of mortality by reducing body weight and condition (Chapter 7). The frequency of disease occurrence also correlates with density (Figures 7.19, 8.13). However, it was not possible to estimate disease-induced mortality (cf. Anderson & May, 1978) and thereby determine whether the disease also acted in a density-dependent manner (Leader-Williams, 1982).

Table 8.2. *Composition counts in three reindeer herds on South Georgia in March 1976*

Herd	N	Calves/100 adults and yearlings	Calves/100 adult females	Adult and yearling males/100 adult and yearling females	Adult males/100 females
Royal Bay	143	36.1	77.5	69.3	57.0
Busen	600	38.2	78.6	57.8	54.0
Barff	739	35.8	63.9	43.5	29.5

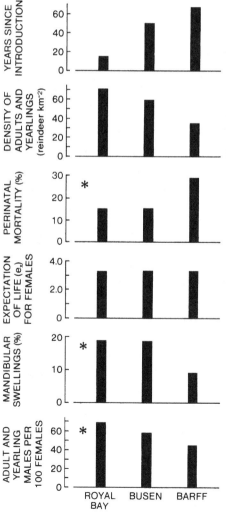

Figure 8.13. The stage of irruption and density of adult and yearling reindeer at the start of the study, in relation to parameters describing mortality and disease in each herd on South Georgia. Significant differences in mortality and disease between herds are shown with * = $P < 0.05$ (all data are derived from tables and figures in earlier chapters).

The densities of adult and yearling reindeer in each herd at the start of the study were derived from data in Table 3.3. Comparative data for the three herds on rates of perinatal mortality, expectation of life for females, the frequency of mandibular swellings and the ratios of adult and yearling males to females were derived as described for the Barff herd in Chapter 7. Chi-square tests show that rates of perinatal mortality differ between herds ($\chi^2 = 7.40$, $df = 2$, $P < 0.025$), but there is a departure from a linear trend in proportions (Snedecor & Cochran, 1969) between perinatal mortality and the density of reindeer in each herd ($\chi^2 = 13.52$, $P < 0.001$). The frequency of mandibular swellings differs between herds ($\chi^2 = 7.44$, $P < 0.025$), but there is no departure from a linear trend in proportions between the frequency of swellings and density ($\chi^2 = 0.23$). Ratios of adult and yearling males and females also differ between herds ($\chi^2 = 6.83$, $P < 0.025$), and again there is no departure from a linear trend between these ratios and density ($\chi^2 = 0.23$).

Little is known of the patterns or rates of mortality of introduced reindeer on other islands, apart from St Matthew (Klein, 1968). There was reduced recuitment prior to the population crash (Figure 8.14) and, as a result of it, there was an exceptionally biased adult sex ratio amongst the survivors, in which only one out of 42 reindeer was male (Figure 8.15). The available data, therefore, suggest that a more drastic limitation of winter forage caused the crash, which provided a speeded-up version of the responses seen on South Georgia.

8.44   *Climate and emigration*

I established earlier that neither latitude (and presumably temperature), nor island size appear to have affected the success or failure of the introductions of reindeer to a wide variety of islands (Figure 8.1). Instead, with the few available data, I have shown that the main difference between the types of decline seen on islands lies in the dependence by reindeer in winter upon more resilient vascular species rather than fragile lichens. However, density-independent factors like emigration or climate might have subtle secondary effects on the nature of declines seen on islands.

Emigration and dispersal is commonly observed amongst continental herds of *Rangifer* (Bergerud, 1980a; Haber & Walters, 1980). However, introduced reindeer on most small islands cannot emigrate. Thus, as St Matthew, St Lawrence, St Paul and Nunivak islands became increasingly overgrazed by increasing numbers of reindeer, the populations were not able to move to new or unoccupied areas. In contrast, reindeer in Iceland were able to cover a large area which, after overgrazing, could presumably recover before again being reoccupied during a second irruption (Figure 8.4). On South Georgia, also, the retreat of a glacier afforded one herd the opportunity to occupy a new area (Figures 3.9, 3.10) and thereby reduce grazing pressure on an overgrazed range. On both Iceland and South Georgia, the rates of decline in reindeer numbers will therefore have been reduced by the opportunity available for emigration.

Snow cover and its effects on the availability of winter forage, especially when that itself is overgrazed, has been recognised as an important cause of population fluctuations amongst *Rangifer* (Vibe, 1967; Gunn *et al.*, 1981; Thing, 1984; Meldgaard, 1986). No introduc-

Figure 8.14. Measuring skeletal remains of reindeer following the crash die-off on St Matthew Island. (Photograph by D.R. Klein.)

tions have been made to islands that are outside *Rangifer*'s overall climatic range, but islands that themselves span over 20° of latitude have different prevailing climates. St Matthew Island has a mean annual temperature of *c.* −2.5 °C compared with that on South Georgia of 2 °C (Klein, 1959; Figure 3.4). Thus the crash on St Matthew Island occurred because of a combination of overgrazed range and an exceptionally severe winter (Klein, 1968). Similarly, the decline of reindeer in Iceland during the 19th century (Figure 8.4) was associated with a combination of overgrazing and severe winters (Thórisson, 1983). In contrast, the milder climate of South Georgia has been warming gradually since the 1930s (Figure 8.16). This warming has coincided with a period when winter forage was plentiful and only latterly decreased in availability. If this warming trend reverses, its effect on the reindeer would now be more serious than during the colder spells of the 1920s, and faster declines could then be expected.

In this section, therefore, I have shown that both climate and the opportunity to emigrate can modify the rates of decline of introduced reindeer on islands. Indeed, the available data might even suggest these factors to be of greater importance than the nature of the winter forage, as I have argued previously. I believe this unlikely, however, because no common trends in latitude or island size can be detected across failed and successful introductions (Figure 8.1). Furthermore, my conclusions are supported by recent findings from naturally-occurring, High Arctic populations of *Rangifer* that have all been shown to depend on a non-lichen diet in winter (Chapter 5), and could easily be tested further by a study of the winter forage of reindeer on, for example, Nunivak or St Lawrence islands.

Figure 8.15. Ratios of reindeer calves, yearlings and males for 100 females on St Matthew Island before and after the population crash during the winter of 1963/4 (data from Klein, 1968).

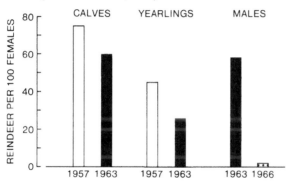

8.5   Irruptions in natural populations

Irruptions have frequently been observed in naturally-occurring populations of *Rangifer* in recent times (Figures 1.8, 1.9). During the course of their history, the various sub-species have adapted to habitats that span over 35° of latitude and range from taiga woodland to High Arctic tundra (Figure 1.2). A variety of limiting factors might have been expected to operate over this range of habitats and sub-species. Data from recent research on *Rangifer* (e.g. Kelsall, 1968; White *et al.*, 1981; Bergerud, 1983; Tyler, 1987a) permits a reconstruction of the probable life history of *Rangifer*, before the influence of modern man modified its natural habitats and population dynamics.

In continental areas, *Rangifer* depended in winter on a lichen-based food niche. They migrated to higher quality summer pastures, and thereby also minimised harassment from biting flies. Rates of fecundity, within the limits imposed by single births, were generally high. However,

Figure 8.16. Climatic data from Grytviken from 1906 to the 1980s. 'Winter' precipitation (both snow and rain) was measured in June, July and August, and length of 'winter' was defined as the number of months in which mean air temperature was below 0 °C (data by courtesy of the British Antarctic Survey, from Leader-Williams, 1980d).

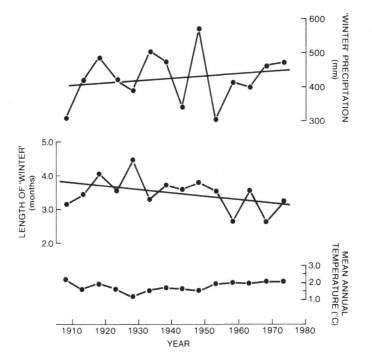

natural predation by wolves and lynxes was directed primarily at juveniles, which thus suffered high rates of mortality. Adult *Rangifer* were also harvested by indigenous peoples using traditional methods. By contrast, the scope for migration on High Arctic islands was limited and *Rangifer* there depended on a non-lichen diet in winter. Large deposits of fat were laid down in summer and fecundity rates varied greatly due to annual differences in plant productivity, that in turn affected recruitment. Natural predators and indigenous hunters were absent or present only in low numbers. There was clearly a gradation in limiting factors from the High Arctic, where food and its effects on fat reserves and fecundity have been of primary importance, to continental areas where predators had a dominant role (Figure 8.17).

Though many of these controls still operate to limit population size, their balance has been disturbed in modern times. The development and widespread use of modern firearms has perhaps been of most importance (Bergerud, 1974c; Bergerud *et al.*, 1984; Miller, 1983). Firearms have permitted two somewhat conflicting factors to come into play. On the one hand, they have permitted the direct overharvesting of *Rangifer* in both continental and High Arctic areas. On the other hand, firearms removed a check in continental populations by permitting the reduction or eradication of natural predators. The modern biologist now working on factors regulating natural populations of *Rangifer* has to disentangle natural from man-induced limitations on their irruptions. In seeking a common answer to the question of what limits continental *Rangifer* populations, a controversy has arisen over whether predation or forage is the most important (e.g. Bergerud, 1983).

But are these two factors mutually exclusive? I believe not. Irruptions of introduced reindeer on islands are natural experiments in which continental sub-species of *Rangifer* have been transplanted to situations more resembling those faced by High Arctic sub-species. These experiments show the ways in which different aspects of forage quality and quantity can affect life history parameters which, in turn, have determined the course of irruptions on islands. Similar factors may well be operating in continental areas to modify rates of change primarily caused by different levels of natural predation or hunting mortality (Figure 8.5).

Caribou and reindeer populations subjected to predation suffer higher rates of calf mortality than non-predated mainland or introduced island populations (Figure 8.18). The importance of predation as a regulating factor need therefore not be disputed and, indeed, disproof of the predation–limitation hypothesis for continental populations is still

Figure 8.17. *Rangifer* populations adapted to different natural limitations, before the development of modern firearms. Continental sub-species have been limited primarily by wolves (A), whilst High Arctic sub-species have been limited by food abundance (B). (Photographs by D.D. Roby and N.J.C. Tyler.)

awaited (Bergerud, 1980a, 1983; Bergerud *et al.*, 1983). However, the importance of differences in range quality may also have been too easily dismissed, as shown by comparisons between two herds of barren-ground caribou (Figure 8.19). Both herds lie within predated cells in Figure 8.5, but had different rates of fecundity and increase at the time they were studied. The declining Kaminuriak population was supposedly limited only by predators, and therefore living well below some aspect of range carrying capacity (cf. Bergerud, 1980a). This gives rise to the expectation of high fecundity rates in both herds, yet these were not observed in the declining herd (Figure 8.19). Recently, however, the reproductive potential of the Kaminuriak herd seems to have increased (Heard & Calef, 1986). Wolves have failed to limit numbers in the more fecund George River herd which continues to expand with a rate of increase = 0.11 (Parker, 1981; Parker & Luttich, 1986).

Studies of introduced island reindeer suggest that an aspect of range quality will underlie differences in fecundity of caribou, yet no data on possible differences in forage quality or productivity (cf. Figures 8.9, 8.10) are available for these populations. Future studies in continental *Rangifer* might thus benefit from seeking to elucidate how food and

Figure 8.18. Variations in the mean ± SE ratios of calves to adult females in late summer amongst *Rangifer*, showing high calf loss due to predation in continental areas, variable rates of fecundity in the High Arctic, and reduced loss of calves amongst continental sub-species introduced to islands (data from: Bergerud, 1980a; Reimers, 1977; Klein, 1968; Thórisson, 1983; Table 8.2).

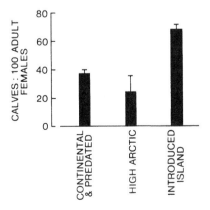

Figure 8.19. Differences in fecundity of two herds of caribou both subjected to natural predation and hunting mortality: the more fecund George River herd continues to expand, whilst the less fecund Kaminuriak herd was declining (data from Dauphiné, 1976; Parker, 1981).

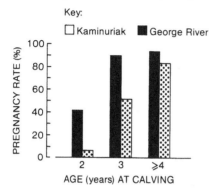

predators combine to limit populations, instead of continuing the exclusive pursuit of one or other hypothesis (cf. Sinclair, 1985).

8.6   Summary

1. Reindeer have been introduced to a wide variety of islands throughout the world. Less than half of these introductions have resulted in established populations. About a quarter of them failed after an initial irruption, and dramatic crashes occurred on some islands.
2. Rates of increase in introduced populations have varied and are determined by quality of summer forage and its effects on female body size. The maximum numbers attained on each island were probably related to the standing crop of available vegetation.
3. Introduced reindeer that have depended on a lichen diet in winter have crashed, whilst those populations that eat more resilient vascular species have shown more gradual declines.
4. As introduced reindeer move through the stages of their irruption, juvenile mortality increases and adult sex ratios become more biased towards females. Emigration and climate have modified population declines on some islands.
5. Comparisons of irruptions suggest that forage may play a secondary role to predators in limiting continental *Rangifer* populations.

9

Influence of reindeer on vegetation

Natural habitats on small islands seem to be much more vulnerable to invading species than those of continents. This is especially so on oceanic islands, which have rather few indigenous species.

Charles Elton (1958)

9.1 Introduction

In the Northern Hemisphere, there is much evidence documenting the effects of grazing by *Rangifer* upon continental ranges, and especially upon the response of lichens (Hustich, 1951; Pegau, 1970a; Andreev, 1977; Helle & Aspi, 1983; Skogland, 1983). However, no evidence has yet emerged to demonstrate that forage plays a role in limiting population numbers where predators are present. In contrast, forage is likely to play a central role in limiting numbers of *Rangifer* on islands lacking predators (Chapter 8). South Georgia is an interesting site in which to study the effects of grazing by reindeer and the possible role of forage in limiting population numbers, for two main reasons. First, the floras of southern islands are not normally grazed by herbivores any larger than beetles, and this has permitted the development of plant communities that are highly sensitive to grazing by mammals (Holdgate & Wace, 1961). Secondly, simple communities comprising few species have little capacity for ecological readjustment (Elton, 1958). As a result, the effects of reindeer upon the vegetation of South Georgia are easily documented (Bonner, 1958; Lindsay, 1973; Kightley & Lewis Smith, 1976; Leader-Williams *et al.*, 1981).

In this chapter I first document the extent of the effect of reindeer on the island's vegetation (9.2). I use the term 'overgrazing' specifically to denote major changes in the density or stature of a species and resultant changes in the composition of native communities. The term is appropriate in the context of southern island floras which are not adapted to any form of grazing by herbivorous mammals. Secondly, I present results from a long-term experiment in which reindeer have been prevented from grazing various plant communities on South Georgia (9.3). This experiment has now run for 12 years since its establishment in 1973/4 (Kightley & Lewis Smith, 1976). The experiment permits an objective assessment of change in communities under continued grazing pressure and of those communities and plant species that recover if grazing pressure is removed.

I discuss the following questions (9.4). First, does overgrazing follow a sequence related to the stage of irruption of each herd? In New Zealand a decrease in the food supply caused a downturn in numbers in the irruption of Himalayan thar (Caughley, 1970, 1976a,b). Secondly, does the sequence of overgrazing on South Georgia follow that normally seen on other southern islands to which herbivorous mammals have been introduced? Tussock grassland is usually the most vulnerable community on southern islands and other communities are generally affected only when grazing pressure is high (Holdgate & Wace, 1961). Thirdly, can an area affected by overgrazing return to its full biological productivity? The

Figure 9.1. Examples of effects of reindeer on the vegetation of South Georgia: (A) and (B) show two tussock grassland sites in Royal Bay with tussocks in different classes ranging from: D (totally denuded) to 1 (vigorous growth); (C) shows an erosion site on the Barff Peninsula in overgrazed tussock grassland.

productivity of such areas may be permanently reduced if removal of vegetation and exposure to wind and water erosion is excessive (Hold-gate & Wace, 1961). Finally, how might the outcome of an irruption be affected by characteristics of the food supply? Reindeer populations introduced to some Arctic islands have crashed, whilst other introductions have had a less dramatic outcome (Chapter 8). I have suggested

factors that underlie these different outcomes, but have not yet substantiated this with any experimental evidence.

9.2   Extent and sequence of overgrazing

Reindeer have occupied three small areas of South Georgia for different lengths of time (Table 3.3). Studies of other irruptions suggest that the sequence of overgrazing in each area will follow a sequence that is related to the stage of irruption (Figure 2.1). Reindeer had already overgrazed macrolichens from dry meadow and fellfield communities in the most recently occupied area within 12 years of their arrival (Lindsay, 1973). Therefore, I examined less fragile components of the vegetation for any trends in overgrazing, using both extensive qualitative surveys and more intensive quantitative methods.

9.21   *Types and sites of overgrazing*

Extensive qualitative surveys of the reindeer areas were completed on foot during 1974 to 1976. I noted and mapped areas of overgrazing and of erosion possibly caused by overgrazing (Figures 9.1 and 9.6). Earlier work had shown that, besides macrolichens, loss of *Poa flabellata*, *Acaena magellanica* and *Deschampsia antarctica*, and the invasion of the introduced *Poa annua* were the most obvious effects of reindeer grazing (Bonner, 1958; Kightley & Lewis Smith, 1976). Selection for these species by reindeer was later confirmed (Chapter 5). Thus I concentrated upon tussock grassland, mesic meadow, dwarf-shrub sward and mossbank communities.

> Extensive surveys were undertaken only after familiarisation with vegetation characteristics in ungrazed areas (Figure 3.5). The following categories were adopted: (1) tussock overgraze: areas of former tussock grassland that had been reduced to bare tussocks or mossbanks (Figure 9.1A); (2) dwarf-shrub overgraze: areas of former dwarf-shrub sward that had been converted to dead litter or grown over by mossbanks through loss of *Acaena magellanica*; (3) invasion of introduced species: mesic meadows or tussock grasslands in which the dominant species had been reduced or eradicated, and replaced by *Poa annua*; (4) erosion sites: areas where vegetation cover had been reduced (as in 1 & 2) but where, in addition, topsoil had slipped extensively (Figure 9.1C). Sites of overgrazing were mapped but are presented here in a less detailed diagrammatic format (Figure 9.2).

Results from the extensive surveys confirm the suggestion of a sequence of overgrazing in the three areas (Figure 9.2). Severe overgrazing was evident in three categories over much of the Barff and Busen areas occupied by reindeer in 1976. In the Royal Bay area, tussock grassland was more extensive (Figure 9.1C), dwarf-shrub sward was still present even though it was composed primarily of woody litter, and *Poa annua* had not yet invaded all mesic meadows. Earthslips in areas where vegetation had been overgrazed were found only on the Barff Peninsula, the area first occupied by reindeer. These sites occurred only on steep north-facing slopes on the east coast of the peninsula (Figures 9.1C, 9.2A). Recently it has been debated whether introduced herbivores cause any more erosion on southern islands than would occur naturally without them. Current evidence from New Zealand and Macquarie Island suggests that introduced herbivores may have been blamed unjustly for this problem (Griffiths, 1979; Adams, 1980; Scott, 1983; Selkirk, Costin, Seppelt & Scott, 1983). There is no firm evidence from South Georgia, but nevertheless I include sites of erosion in the present survey in case future work shows grazing by reindeer to be an important causal factor.

9.22 *Patterns of tussock degradation*

Tussock grassland is highly productive and is the only community available to reindeer in winter (Chapter 5). Therefore, it is probably an important determinant of carrying capacity and of the upper limit of reindeer numbers on South Georgia (Chapter 8). Hence, more intensive surveys of the extent of overgrazing in this community were undertaken to assess its role in the irruptive sequence.

More intensive surveys of tussock grassland in different areas were undertaken during the summers of 1974/5 and 1975/6. A total of 15 rectangular transects of varying size, dependant on topography, were laid out in each area, including ungrazed areas of the Busen Peninsula (Table 9.1). Each transect was categorised according to physiographic site into: (1) raised beach; (2) ridgetop; (3) slope accumulating drift snow (= drift slope). Total numbers of tussock pedestals were counted in each transect, and each tussock was assigned to one of five classes: (D) a totally denuded pedestal comprising a rounded bump, with or without moss; (4) a pedestal with mostly dead leaves; (3–1) tussocks of increasing size and vigour. Examples of tussocks D–1 are shown in Figures 9.1A and B (other examples of 2 & 1 in ungrazed areas are shown in Figure 3.5A). For each transect the proportion of tussocks in each class was expressed as $p_D + p_4 + ...p_1 = 1$. A damage score (*DS*) was then calculated as follows: $DS = (4 \times p_D + 3 \times p_4 + 2 \times p_3 + 1 \times p_2 + 0 \times p_1)$. The *DS* resulted in a maximum of 4 for a transect with totally denuded stools, and a minimum of 0 for a transect comprising only tussocks in class (1). Differences in *DS* were compared across topographic sites and reindeer areas using a Kruskal-Wallis ANOVA or a Mann-Whitney *U* test with two-tailed probability values (Siegel, 1956).

Figure 9.2. Diagrammatic representation of extent of overgrazing on South Georgia in 1976, shown in the sequence of occupation of each area with (A) Barff (1911); (B) Busen (1925); (C) Royal Bay (1961–5).

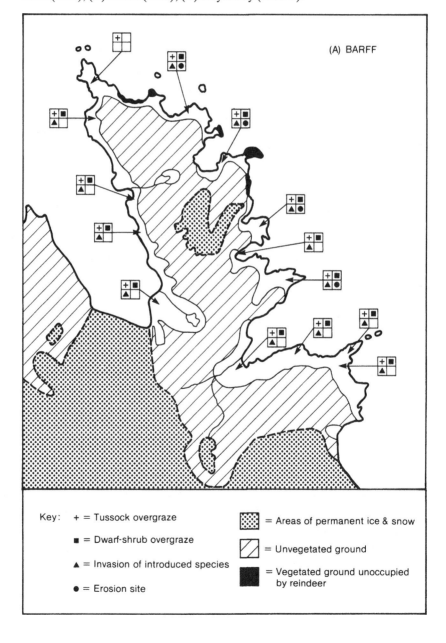

The proportions of tussocks in each class vary in the different reindeer areas and topographic sites (Table 9.1, Figure 9.3). The areas of the Busen Peninsula then unoccupied by reindeer (Figure 9.2B) serve as a control, and show that ungrazed tussocks occur in a range of size classes which presumably reflect population dynamics of the community under natural conditions. In areas that have been occupied by reindeer for successively longer periods, there is an increasing proportion of tussocks

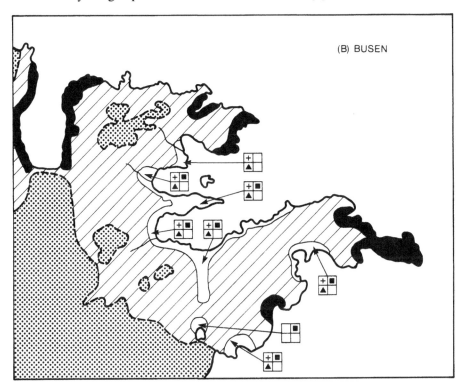

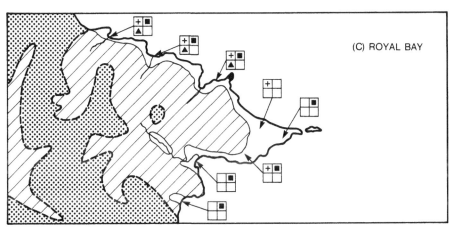

in more damaged classes in ridgetop and raised beach sites. However, tussocks on drift slopes are less affected by damage from grazing.

Damage scores calculated from proportions of tussocks in each class provide an index of differences between areas and site types (Figure 9.4). Scores differ across topographic sites within each reindeer area. Scores also differ for raised beach and ridgetop sites between areas. However, ridgetop sites have higher damage scores sooner after occupation by reindeer than raised beach sites. In contrast, scores for drift slope sites do not differ between areas. This suggests a sequence of overgrazing, in which tussocks on ridgetops are the first to be damaged, followed by those on raised beaches. Drift slopes are better protected by deeper snows and remain the least overgrazed sites. When exposed by snow melt, drift slopes in summer still provide extensive areas of relatively undamaged tussock grassland, even in the area occupied for longest by reindeer.

The survey shows that overgrazing of tussock grassland follows a

Table 9.1. *Summary of survey data from tussock grassland sites in each reindeer area. Derivation of the damage score is explained in the text, and results are shown as mean $\pm SE$*

Area	No. of transects	No. of tussocks	Total area (m^2)	Tussock density	Damage score
Barff					
Raised beach	6	139±37	175±43	0.78±0.08	3.45±0.26
Ridgetop	5	304±77	470±99	0.66±0.10	3.27±0.13
Drift slope	4	224±105	314±122	0.74±0.13	1.64±0.28
Total	15	3247	4655	0.70	
Busen					
Raised beach	4	155±41	165±41	0.93±0.08	2.73±0.15
Ridgetop	5	114±17	165±15	0.68±0.05	3.28±0.11
Total	9	1192	1485	0.80	
Royal Bay					
Raised beach	5	163±26	197±36	0.85±0.06	1.64±0.22
Ridgetop	5	127±22	147±17	0.83±0.08	3.03±0.16
Drift slope	5	147±12	201±17	0.74±0.05	1.27±0.13
Total	15	2182	2725	0.80	
Busen ungrazed					
Raised beach	3	173±26	190±38	0.93±0.05	0.55±0.04
Ridgetop	3	93±33	123±35	0.73±0.07	0.85±0.04
Total	6	799	940	0.85	

sequence that is related to the length of time each area has been occupied by reindeer (Figure 9.4). This supports suggestions made earlier (Chapter 8) that a decrease in the food supply is a causal factor in the downturn of reindeer numbers in the irruptive sequence on South Georgia.

Kruskal-Wallis ANOVA shows an overall difference in damage scores between different physiographic sites and reindeer areas ($H = 37.71$, $df = 9$, $P < 0.001$). Kruskal-Wallis ANOVA and Mann-Whitney U tests show there are differences in scores by sites within areas (Barff: $H =$

Figure 9.3. Proportion of tussocks in each class, shown as mean ± SE, in different topographic sites in each reindeer area. Classes varied from: D (totally denuded pedestal) to 1 (vigorous growth).

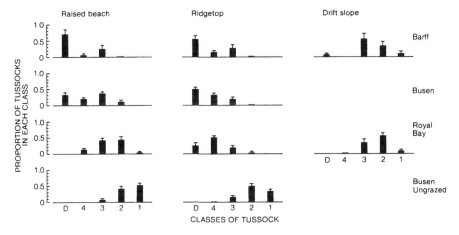

Figure 9.4. Damage scores, shown as mean ± SE, in tussock grassland in different topographic sites in each reindeer area (A) and at different times since occupation of each area (B). Significance levels shown with ** = $P < 0.01$, * = $P < 0.05$.

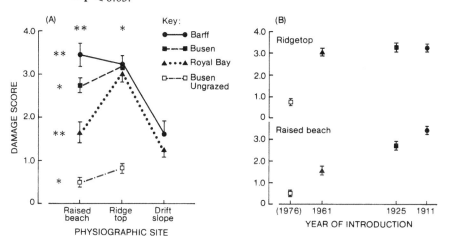

8.90, n_j = 6, 5 & 4, $P < 0.01$; Royal Bay: H = 10.10, n_j = 5, 5 & 5, $P <$ 0.01; Busen: U = 1, n_1 = 4, n_2 = 5, $P < 0.05$; Busen ungrazed: U = 0, n_1 = 3, n_2 = 3, $P < 0.05$). There are differences in scores by areas within raised beaches (H = 14.62, df = 3, $P < 0.01$) and ridgetops (H = 8.62, n_j = 5, 5, 5 & 3, $P < 0.05$). However, there are no differences in scores between the three ridgetop sites occupied by reindeer (H = 2.16, n_j = 5, 5, & 5, $P > 0.10$) nor between drift slope sites (U = 5, n_1 = 4, n_2 = 5, $P > 0.10$).

9.23   *Sequence of impact on the vegetation*

On southern islands tussock grassland is generally the vegetation type most vulnerable to introduced herbivores, whilst other communities (e.g. fernbush, broad-leaved herbfield and fellfield) have been affected severely only when grazing pressure is high (Holdgate & Wace, 1961). On South Georgia, however, fellfield, dry meadow, dwarf-shrub sward and mesic meadow communities were affected first by reindeer after introduction or emigration to a new area (Bonner, 1958; Lindsay, 1973; Kightley & Lewis Smith, 1976). Tussock grassland was affected severely only when reindeer had achieved their peak densities, contrary to generalisations made for other southern islands (Holdgate & Wace, 1961).

Three factors explain this difference on South Georgia. First, reindeer have a predeliction for macrolichens (Chapter 5) which occur mainly in dry meadow and fellfield communities. These were soon overgrazed on South Georgia, as they have been in areas of the Arctic subjected to continuous grazing by non-migratory reindeer (Lindsay, 1973). No other species of mammal introduced to southern islands (Figure 2.7) selects for lichens (Chapter 10). Secondly, reindeer have overgrazed preferred vascular species available in summer (*D. antarctica* and *A. magellanica*) from a flora in which species low in macronutrients predominate. As a result, dwarf-shrub sward is largely unrecognisable as a community in reindeer areas and mesic meadows are invaded by *P. annua*. The forage preferences of other species of introduced herbivores on southern islands are largely unknown, but most likely differ from reindeer on South Georgia (Chapter 10). Thirdly, snow on other subantarctic islands with introduced herbivores only lies deeply and for short periods at higher altitudes (Walton, 1984). Whilst areas of tussock grassland always remain available during winter on South Georgia (Chapter 5), snow cover protects extensive areas of this community in drift slope sites (Figure 9.4). Therefore, only the most accessible sites on ridgetops and raised beaches have been overgrazed, in contrast to other southern islands where all tussock grassland sites are equally prone to overgrazing by introduced herbivores (Holdgate & Wace, 1961).

Additionally, snow cover on South Georgia causes a large seasonal reduction in carrying capacity. Thus reindeer may not have achieved the densities which would have been possible in the absence of snow cover, and grazing pressure on tussock grassland has been lower than its absolute standing crop would otherwise have allowed. Reindeer have been introduced to Iles Kerguelen (Figure 2.7) where, due to less heavy and frequent snowfall, the relative reduction in carrying capacity between summer and winter is less marked. A comparative study of the maximum densities attained by these reindeer, their grazing preferences and the sequence of overgrazing would be of interest in relation to that seen on other southern islands (Holdgate & Wace, 1961).

9.3   A long-term exclusion experiment

The impact of herbivores upon vegetation is frequently assessed by excluding herbivores and permitting the vegetation to grow in the absence of foraging pressure. Exclusion experiments have been used in many studies of herbivore–vegetation interactions, including the assessments of recovery of vegetation grazed by reindeer and caribou in the Northern Hemisphere (Palmer & Rouse, 1945; Pegau, 1970b, 1975). Such an experiment was established on South Georgia (Kightley & Lewis Smith, 1976) with the aim of showing which communities and plant species on South Georgia respond to exclusion from grazing (Leader-Williams, Lewis Smith & Rothery, 1987).

Nineteen experimental sites were established in six different community types in the three reindeer areas (Table 9.2, Figure 9.5) in 1973/4, as described by Kightley & Lewis Smith (1976). The sites comprised a total of eight exclosures (10 × 10 m) and 11 cages (2.5 × 2.5 m). Adjacent to each fenced exclusion plot, an unfenced control plot of the same dimensions remains freely accessible to the reindeer. Permanent quadrats (10 by 1 m² in tussock grassland and dry meadow exclosures; 15 by 0.25 m² in other exclosures; 2 by 0.25 m² in cages) are marked out within each exclusion and control plot. One exclosure was damaged irreparably by elephant seals in 1974 and was not maintained. In the remaining 18 experimental sites, vegetation cover has been monitored once per year in January and February, as often as possible between 1973/4 and 1985/6, for a total of eight to ten seasons per site. All phanerogamic and cryptogamic species occurring in each quadrat were listed and a visual estimate of the percentage cover afforded by each and by bare ground was determined to the nearest 5%: lesser proportions were recorded as trace. As six different observers undertook these recordings over 12 years, photographs of each quadrat were taken as an additional record each year.

Table 9.2. *Location and community type of exclosures (E) and cages (C) monitored during 1973 to 1986: experimental sites are shown in Figure 9.5 (after Kightley & Lewis Smith, 1976)*

	Area		
Community	Barff	Busen	Royal Bay
Tussock grassland	(2E)	16E	10E
Mesic meadow	9C	19C	12C
Mossbank	4C, 6C	18C	
Dry meadow/dwarf-shrub sward	3E, 5C, 7C	15E	11E
Oligotrophic mire	1E, 8E	17C	13C
Eutrophic mire			14C

Figure 9.5. Sites of experimental exclosures and cages in the reindeer areas of South Georgia (from Leader-Williams *et al.*, 1987).

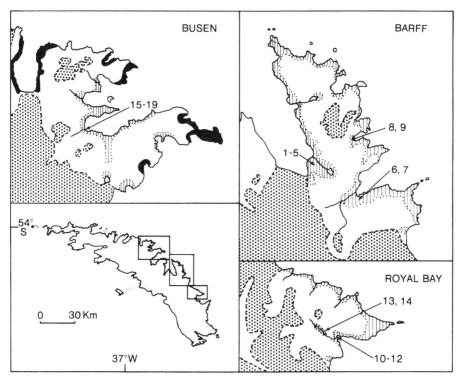

In the analysis, cover values of the following taxa were combined into single variables: both *Colobanthus* spp.; *Polytrichum* spp. and *Chorisodontium aciphyllum*; *Calliergon sarmentosum* and *Drepanocladus uncinatus*; *Pohlia wahlenbergii* and *Philonotis acicularis*; all other mosses; all liverworts; all *Cladonia* spp.; all other lichens; all algae; all macrofungi. Data for cover values of each species or species combination were examined for the following: (1) initial difference between exclusion and control plots at establishment; (2) trends with time in exclusion and control plots; (3) differences in trend between exclusion and control plots and (4) similarities or differences in response between each species occurring in exclusion and control plots. Student's *t* test was used to compare initial differences, after angular transformation of cover values. All other trends and responses were examined using the Kendall rank correlation coefficient, as described by Leader-Williams *et al.* (1987).

9.31 Start of the experiment

Nineteen experimental sites were established in six different community types in the three reindeer areas (Table 9.2, Figure 9.5). Sites comprised fenced exclusion plots, either larger exclosures or smaller cages, and control plots that remained freely accessible to reindeer (Figure 9.6). A total of 29 species (or species groups) and bare ground were recorded in the experimental sites. Nineteen of these species occurred either infrequently or at low cover values throughout the experiment (Table 9.3) and will not be considered in as much detail as the remaining species. Comparisons of cover values of each species occurring in exclosures and in their respective control plots at the start of the experiment showed differences in only 11% of possible cases ($N = 103$). Thus, each exclosure and control plot were well matched with respect to cover values for most species at the start of the experiment. The destruction of exclosure 2 by elephant seals in its first season, and the realisation that the site could never be protected from seal harems during the breeding season, unfortunately eliminated the most degraded coastal tussock grassland site from continued monitoring.

Comparison of mean cover values for each species inside exclosures and in controls at the start of the experiment with Student's *t* test showed few differences. In 103 possible comparisons, only three pairs differed at $P < 0.01$ (*D. antarctica* in site 1 with a mean cover of 6% in the exclosure and 13% cover in the control plot; *A. magellanica* in site 11 with mean cover values of 63% vs. 43%; *Tortula robusta* in site 1 with mean cover values of 98% vs. 64%) and only eight pairs differed at $P <$

Figure 9.6. Changes in vegetation cover over time at selected experimental sites: (A) and (B) show exclosure 10 in January 1974 and February 1979. Tussock grassland site showing continued loss of *Poa flabellata* and an increase in bare ground outside the exclosure: *P. flabellata* has recovered inside the exclosure.

(C) and (D) show enclosure 11 in January 1974 and January 1985. Dry meadow site showing continued loss of *Acaena magellanica* and increase in *Festuca contracta* outside the exclosure: *A. magellanica* has recovered inside the exclosure.

Figure 9.6 (cont.)
(E) shows cage 9 in February 1979. Mesic meadow site showing increased presence of *Poa annua* outside the cage: native species, notably *Deschampsia antarctica*, have recovered inside the cage. (F) shows cage 18 in January 1982. Moss bank site showing no recovery of vascular species. (Photographs by courtesy of British Antarctic Survey.)

Table 9.3. *Frequency of occurrence of less important species and species groups in eighteen experimental sites shown for both trace and larger cover values (from Leader-Williams et al., 1987)*

Species and species groups	Tussock grassland N = 2		Mesic meadow 3		Mossbank 3		Dry meadow/ dwarf-shrub sward 5		Oligotrophic mire 4		Eutrophic mire 1		Total (%) 18	
	≥5%	T	≥5%	T	≥5%	T	≥5%	T	≥5%	T	≥5%	T	≥5%	T
Alopecurus magellanicus			1										1(6)	
Poa pratensis[a]										1				1(6)
Uncinia meridensis								2		1				3(17)
Juncus scheuchzerioides			1					2	2	2			3(17)	4(22)
Acaena tenera	1	1					2	2		1			3(17)	4(22)
Ranunculus biternatus			1				2						3(17)	
Colobanthus spp.			1					2		2			1(6)	4(22)
Montia fontana			1	1				1					1(6)	2(11)
Galium antarcticum										1				1(6)
Callitriche antarctica		1		1						2				4(22)
Cerastium fontanum[a]	1			1				1					1(6)	2(11)
Taraxacum officinale[a]								1						1(6)
Calliergon & Drepanocladus spp.			1						2	2	1		4(22)	2(11)
Pohlia & Philonotis spp.									1	1	1		2(11)	1(6)
Other mosses		1		2				2	2			1	2(11)	6(33)
Liverworts	1	1	1		2	1	2	2	2	2		1	8(44)	7(39)
Other lichens		1				1		4						6(33)
Algae								2				1		3(17)
Macrofungi		2		1				1		4		1		9(50)

[a] Introduced species.

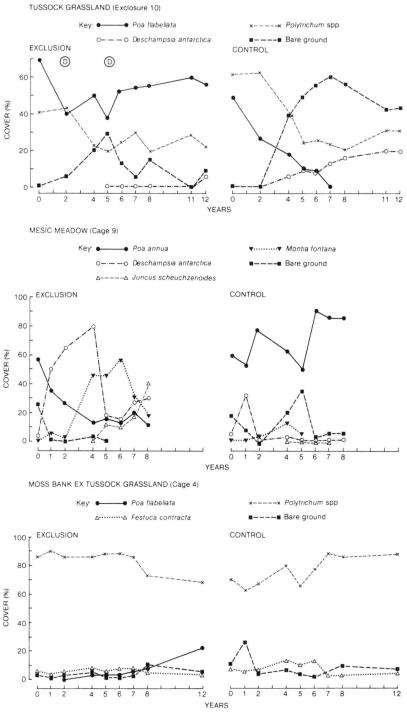

Figure 9.7. Changes in cover values over time in exclusion and control plots in various plant communities. To simplify the interpretation of graphs, only selected species present in each site have been plotted. Years when exclusion

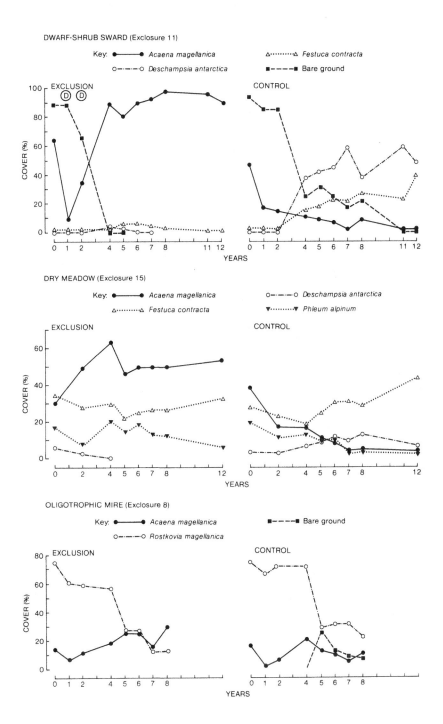

plots were damaged and grazed by reindeer are shown with (D) (from Leader-Williams *et al.*, 1987).

0.05. With no evidence of differences in cover values in 89% of the possible comparisons, it was assumed that exclosure sites and their controls were well matched. This assumption was also made for the smaller cages, though no statistical tests were undertaken because these sites contained only two quadrats per plot.

9.32   *Changes in plant cover over time*

Changes in vegetation cover within many experimental sites have been dramatic during the course of 12 years and, in a few instances, changes occurred after only 2 or 3 years. Considerable changes have occurred in most tussock grassland, mesic meadow and dry meadow sites (Figures 9.6, 9.7). In contrast, changes in mossbank, oligotrophic mire and eutrophic mire sites have been less apparent. Certain of the exclosures and cages were damaged and grazed by reindeer during the course of the experiment, but repaired at the time of the annual recording of data. Such instances of damage served, fortuitously, to validate results from the exclusion experiments. Trends of increase or decrease were reversed during the episodes of damage and were then re-established after repair and continued exclusion (Figure 9.7).

The analysis aimed to identify species that respond to exclusion from grazing and, in particular, to identify consistent differences in behaviour between exclusion and control plots. Thus we looked, first, for any trends of increase or decrease in either the exclusion or control plots and, second, for any differences in the trends of increase or decrease between exclusion and control plots. This analysis identifies many varied and complex changes in cover values for particular species in different sites (Figure 9.8). Differences in response between control and exclosure plots were found in several sites for *P. flabellata*, *D. antarctica*, *A. magellanica* and *Polytrichum* spp., and in at least one site for all but one of the other important species. The exception was *Cladonia* lichens whose low cover values changed so little during the experiment that they are not shown in Figure 9.8. 'Reindeer' lichens have extremely slow growth rates on South Georgia (Lindsay, 1975), and their recovery can be expected to take several decades, as in the Northern Hemisphere (Palmer & Rouse, 1945; Pegau, 1970b, 1975).

In order to identify those species that respond to exclusion most consistently, mean values of the correlation coefficients from all sites were calculated for both exclusion and control plots (Figure 9.9). Only *P. flabellata* and *A. magellanica* show a consistent increase in cover upon exclusion and a decrease in response to continued grazing. In contrast, *P.*

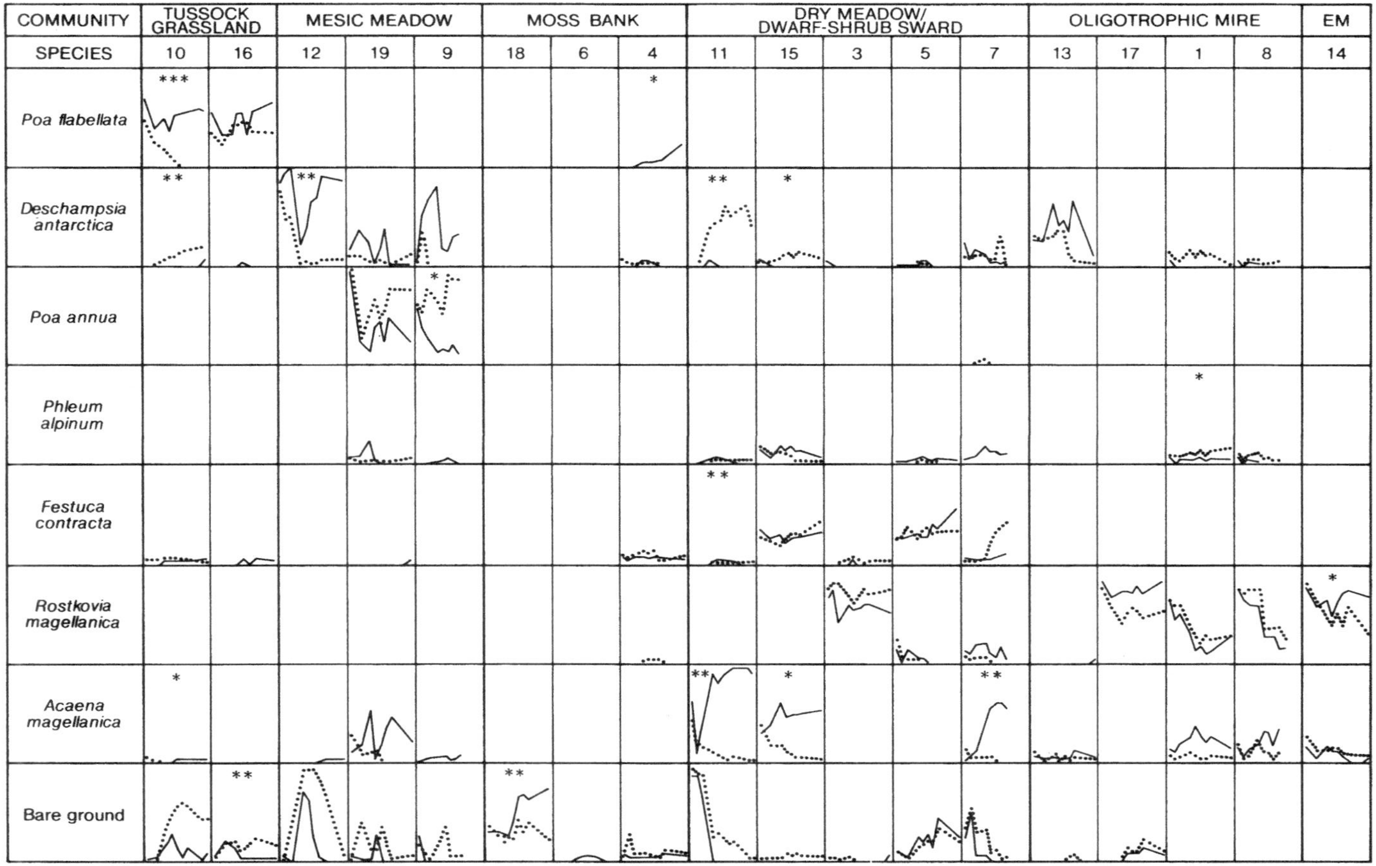

Figure 9.8. Changes in cover values over time for selected vascular species and for bare ground shown for all sites in which they occur. Solid line = exclusion plot; dotted line = control plot. Each box is drawn to scale with vertical axes representing cover values from 0 to 100% and horizontal axes representing time from 0 to 12 years. Differences in the trends of increase or decrease between exclusion and control plots are shown with * = $P < 0.05$; ** = $P < 0.01$; *** = $P < 0.001$ (from Leader-Williams et al., 1987).

annua, *Polytrichum* spp. and bare ground generally decrease in cover upon exclusion and increase in response to continued grazing. The remaining species show less clear and consistent responses in this experiment (Figure 9.9).

Mean cover values for exclusion and control plots were examined for trends over time using the Kendall rank correlation coefficient (Siegel, 1956). Positive values of τ show an increase with time and negative values show a decrease (Leader-Williams *et al.*, 1987). Two-tailed significance levels were obtained from Hollander & Wolfe (1973). The same procedure was applied to differences in cover values between exclusion and control plots to test for any differences in response over time (Figure 9.8). The data were examined for any differences in response of a species in the larger exclosures compared with the smaller cages. No consistent differences were apparent that could be ascribed to the size of the exclusion plot (Leader-Williams *et al.*, 1987). Therefore, correlation coefficients derived from both types of structure were not distinguished in the description of response to exclusion of each species (Figure 9.9).

Figure 9.9. Mean value ± SE of Kendall rank correlation coefficients for important species in all sites in which they occur with cover values >5% (see in Figure 9.8). Solid bar = exclusion plot; shaded bar = control plot (from Leader-Williams *et al.*, 1987).

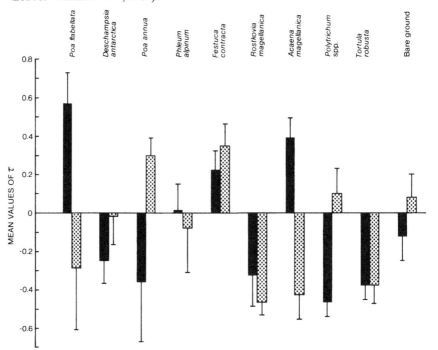

Results from the experiments permit several conclusions. First, only two native species, the grass *P. flabellata* and the dwarf shrub *A. magellanica*, show consistent increases in cover in all the communities in which each occurs if grazing pressure by reindeer is removed. Second, a grazing-tolerant introduced grass (*P. annua*) decreases in cover upon exclusion and invades suitable habitats that are subjected to continued grazing pressure. Third, increases in cover of bare ground and *Polytrichum/ Chorisodontium* mossbanks occur with continued grazing pressure. Fourth, the grasses *D. antarctica* and *F. contracta* show trends of change that depend on the response of other species. The remaining native species of grass, the single sedge, the rushes, forbs, lichens, liverworts, algae and remaining species of moss show either no consistent or no important responses to removal of grazing pressure.

Studies of the South Georgia vegetation growing under natural conditions and ungrazed by reindeer (Lewis Smith & Walton, 1975; Lewis Smith, 1984) have shown differences in the propagative behaviour of different plant groups and species (Table 9.4) and provide support for the conclusions reached in exclusion experiments (Leader-Williams *et al.*, 1987). Grasses generally have high propagative potential through seed production and tillers. The changes found in those species selected by reindeer were much as expected. Rushes also have a high propagative potential, but they are not highly preferred by reindeer and have shown no consistent changes. Of the dwarf shrubs, *A. tenera* grows slowly, produces few rhizomes and has a low seed production. The endemic hybrid, *A. tenera* × *A. magellanica*, is frequently found in some areas of South Georgia and appears to spread only vegetatively. As expected, cover of *A. magellanica* has changed most quickly and dramatically. Forbs generally have a low propagative potential and, as expected, have shown few changes in the experiment. In spite of their overall propagative potential, no vascular species has colonised *Polytrichum/ Chorisodontium* mossbanks (Figure 9.6F). (Only where a vascular species was still present in site 4 has it increased in cover when protected from grazing.) Experimental evidence shows that seedlings can germinate on mossbanks but are unable to obtain sufficient moisture to continue growth (Lewis Smith, 1971). This finding substantiates our exclusion experiment and suggests that mossbanks may only be colonised by vascular species over long periods or not at all.

Table 9.4. *General patterns of propagative behaviour of principal species of vascular plants on South Georgia (from Leader-Williams et al., 1987)*

Group/species	Seed production[a]	Rhizomatous or stoloniferous growth[a]	Principal means of local colonisation[b]	Principal means of distant colonisation[b]	Seedling growth potential on disturbed ground[a]
Grasses					
Poa flabellata	+++		S	S	+
Deschampsia antarctica	++		S & V	S & V	++
Poa annua	+++		S	S	+++
Phleum alpinum	++	+	S & V	S & V	+
Festuca contracta	++	+	S	V	
Rushes					
Juncus scheuchzerioides	+	+++	V	?S	
Rostkovia magellanica	++	+++	V	?S	
Dwarf shrubs					
Acaena magellanica	+++	+++	S & V	S	+
Acaena tenera	++		S	S	
Forbs					
Colobanthus quitensis	++		S	S	+
Ranunculus biternatus	+	+	V	S	
Montia fontana	+		S	S	

[a] Crosses denote importance.
[b] S = seed; V = vegetative; ?S = probably seed, but lacking evidence.

9.33   Differences between areas

Each reindeer area was overgrazed to a different extent at the start of this study (Figures 9.2, 9.4). The same differences in the extent of overgrazing were apparent from analysis of cover values in different communities at the start of the exclusion experiment (Figure 9.10). In the Royal Bay area, the tussock grassland site had a high cover of *Poa flabellata* and dwarf-shrub sward/dry meadow had a high cover of *Acaena magellanica*. There was an inverse relationship between these species and other components of their respective communities. In contrast, the Barff area provided the most degraded sites and the Busen area was intermediate (Figure 9.10).

These differences between areas allowed the following predictions. Native species that respond to exclusion should show the greatest recovery within exclusion plots in the most overgrazed areas. However, within controls they should show the greatest continued loss in the least overgrazed areas. Grazing-tolerant introduced species should show the opposite trend. Thus within exclusion sites, greatest loss should occur in the most overgrazed areas. Allowing for the small sample size of sites in different communities in each area, these predictions were generally borne out over the course of the experiments (Figure 9.11). As with earlier analyses (Figure 9.9), other species showed no clear trends of loss or recovery. Therefore, were grazing pressure by reindeer ever to be removed entirely, these results suggest that ensuing changes in vegetation cover would be more marked in the Barff than the Royal Bay area.

Figure 9.10. Initial cover values of selected species in various communities. Data are shown for experimental sites in different reindeer areas (from Leader-Williams *et al.*, 1987).

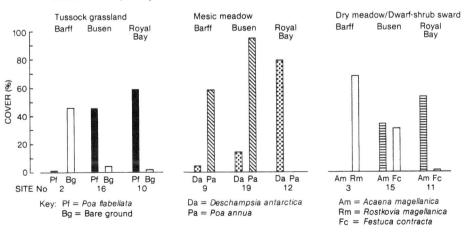

9.4   Irruptions and overgrazing

Changes in forage availability influence the irruptive sequence in ungulate populations. In New Zealand, Africa and North America irruptions can follow upon a dramatic increase in forage availability, whilst a downturn in numbers can follow upon a decrease in food supply (Caughley, 1970, 1976a,b; Sinclair, 1975, 1977, 1979; Houston, 1982; Pellew, 1983). Therefore, I concentrated upon availability of tussock grass, which is selected by reindeer in summer and which predominates in the winter diet. The reduction in the availability of viable tussock pedestals is negatively related to the length of occupation of each reindeer area (Figure 9.4). I consider that this was of major importance in causing the downturn in reindeer numbers during their irruption on South Georgia, and is a result that was expected from earlier studies. However, the factor underlying differences in the final outcome of irruptions on small islands in the Arctic and subantarctic (Chapter 8) still remains to be substantiated.

The characteristics of the winter food eaten by reindeer on St Paul and St Matthew islands and on South Georgia differ markedly. Lichens dominated the winter diet on the two Arctic islands (Scheffer, 1951; Klein, 1968). As a group, lichens are generally of low biomass, of great

Figure 9.11.Changes in cover values of selected species over time in different reindeer areas. Results are shown as mean ± SE (where appropriate) of Kendall rank correlation coefficients. Solid bar = exclusion plot; shaded bar = control plot (from Leader-Williams *et al.*, 1987).

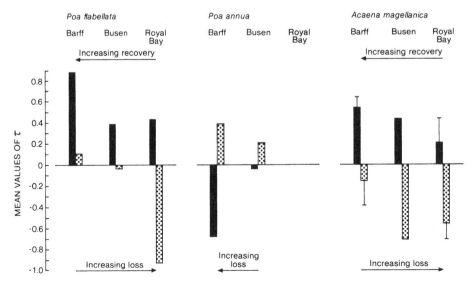

longevity, quickly depleted when subjected to continuous grazing and very slow to recover former abundance when grazing pressure is removed (Palmer & Rouse, 1945; Pegau, 1970b). In contrast, surveys and the exclusion experiment show that tussock grass on South Georgia declines in cover more slowly when grazed and can recover far more quickly than the several decades required by lichens (Figure 9.12). This provides experimental evidence to substantiate earlier assertions that it is the greater resilience of tussock grass compared with lichens which has determined the different outcome of irruptions on small islands (Chapter 8). Future studies of reindeer on other islands would further illustrate the importance of characteristics of their food supply in determining the outcome of an irruption of ungulates (Caughley, 1976b; Laws, 1981).

Once an irruptive oscillation has reached post-decline stages, it has been suggested that the original composition of the vegetation is likely to be altered irreversibly. Fears have been expressed about possible extinctions of plant species and about severe erosion (Holdgate & Wace, 1961; Riney, 1964). In contrast, a more recent review has provided no evidence of any irreversible consequences arising from the many introductions of ungulates to New Zealand (Caughley, 1981). On South Georgia no species of vascular plant is known to have been lost from areas occupied by reindeer for several decades (Greene, 1964), even though the

Figure 9.12. Model of response of lichens on St Matthew and tussock grass on South Georgia to introduction of reindeer and to removal of grazing pressure. The recovery phase was initiated by the crash on St Matthew and by the exclusion experiment on South Georgia (based on data from Klein, 1968 and Figures 9.4, 9.8; from Leader-Williams *et al.*, 1987).

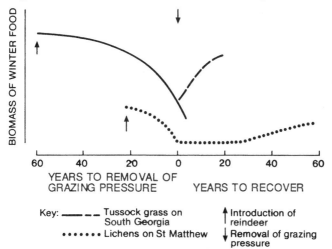

composition of plant communities differs radically from their ungrazed state. As both tussock grass and dwarf shrubs can recover when protected from grazing, even at the expense of an introduced grazing-tolerant species, many of the more important components of the vegetation affected by reindeer could probably regain their former abundance. Indeed, recovery of overgrazed vegetation has occurred previously on South Georgia on a small offshore island formerly occupied by sheep (Matthews, 1931; Headland, 1984). Therefore, fears about irreversible change on South Georgia appear to be largely unfounded (cf. Caughley, 1981).

There are, however, two possible exceptions. The first is erosion sites (Figure 9.1C) whose cause and course of recovery upon exclusion it would be desirable to study. The second exception is mossbank communities which have shown little change during the exclusion experiment (Figure 9.6F). If mossbanks cannot become colonised by vascular species, then the reindeer areas will have lost some of their original productivity and carrying capacity. These areas could only support further irruptions of a dampened nature in the future and would not recover fully if grazing pressure were ever to be removed entirely.

9.5   Summary

1. Reindeer have overgrazed tussock grassland, mesic meadow, dwarf-shrub sward and dry meadow communities on South Georgia. Overgrazing may also have resulted in sites of erosion.

2. A sequence in overgrazing of tussock grassland is evident, both in terms of physiographic site and of area. Ridgetop and raised beach sites are the most affected, whilst drift slope sites are the least affected. An increase in damaged tussocks at overgrazed sites follows a similar sequence to the irruptive stage of each herd.

3. Recovery of vegetation in exclosures shows that *Poa flabellata* and *Acaena magellanica* increase in cover in response to removal of grazing pressure. In contrast, the introduced *Poa annua* and, to a lesser extent, *Polytrichum* moss banks and bare ground decrease in cover in response to exclusion. Other species and various communities including moss banks, oligotrophic and eutrophic mires show little change.

4. Tussock grass is depleted slowly when grazed by reindeer and recovers more quickly than lichens. This important difference in

response of the winter food supply to grazing underlies the different outcomes of introductions made to Arctic and subantarctic islands.

5. Few apparently irreversible changes in vegetation composition have occurred on South Georgia as a result of grazing by reindeer.

10

Introduced mammals on southern islands

The fate of remote islands is rather melancholy . . . The reconstitution of their vegetation and fauna into a balanced network of species will take a great many years.
Charles Elton (1958)

10.1　Introduction

Islands hold a fascination for biologists. Studies of island biogeography have played a major role in development of evolutionary theory since the time of Darwin and Wallace. This interest has been maintained by modern biologists, and island biogeography still contributes much to biological theory (e.g. MacArthur & Wilson, 1967; Lack, 1976; Williamson, 1981). However, the faunas and floras of most islands have now been disturbed by man and his activities (Elton, 1958). Exotic species have often been introduced to islands (e.g. Lever, 1985). The impact and importance of such introductions depends upon the fragility and uniqueness of the native floras and faunas. Thus effects of introductions to island groups such as the Galapagos or Seychelles cause concern to conservationists (Stoddart, 1981; Hoeck, 1984).

The ring of southern islands that encircle the Antarctic, and the Antarctic continent itself, are amongst the terrestrial habitats that have been least affected by the activities of man. The remoteness of the islands and the continent, the limited areas available for colonisation and the cold climate have all tended to limit diversity in native floras and faunas. Their degree of endemism is also high. As a result, southern islands provide communities that are relatively simple and of great biological interest for basic studies of biogeography, evolution and adaptation. Therefore, the need to conserve the continent and its surrounding southern islands is being increasingly recognised (Bonner, 1984b; Laws,

1985). Knowledge of the impact of introduced mammals upon simple ecosystems, lacking almost any native land vertebrates, is of importance in achieving this goal (Holdgate & Wace, 1961; Bonner, 1984a; Clark & Dingwall, 1985; Leader-Williams, 1985).

In this chapter I review the characteristics of southern islands and the introductions made to them (10.2.), some of the factors of importance in the establishment of introductions (10.3) and the effects of introduced mammals on native floras and faunas (10.4). Clearly, man himself is amongst the mammals that have been introduced to southern islands. However, I exclude consideration of man's direct effects upon faunas of southern islands which have been well reviewed elsewhere (e.g. Bonner, 1982; Laws, 1985). Instead I concentrate only upon the introduction by man of other mammals. I use the data reviewed in the various sections to discuss the following questions. First, what factors determine whether or not introductions fail or succeed on southern islands? It has often been noted that many introductions fail to become established (de Vos *et al.*, 1956) and I examine possible reasons for this on southern islands. Secondly, are there any common principles that can be observed amongst introductions as diverse as those of cats and reindeer? To date the irruptive model has been discussed largely for ungulates (Caughley, 1970, 1976a,b) and I examine whether it is useful for other species. Thirdly, are the effects of introduced animals upon native biotas largely irreversible? Generally a pessimistic view has been taken of the powers of recovery of island biotas (e.g. Elton, 1958; Holdgate & Wace, 1961).

10.2   Islands and introductions

Insulantarctica is a recently proposed biogeographic realm composed of islands lying between the Antarctic continent and the Subtropical Convergence (Clark & Dingwall, 1985). In this section I first review some of the physical and biological characteristics of 18 islands or island groups that lie within Insulantarctica to the north of 60°S. Second I review the range of human activities and mammalian introductions that have been associated with these islands.

10.21   *Characteristics of southern islands*

Southern islands are all small and remote from major continental landmasses (Figure 2.7). However, they vary in size, elevation, structure and climate (Table 10.1). The Falkland Islands are the largest island group whilst the small islands include Bounty, McDonald and Snares islands. The islands may be classified as either a group, a single main

Table 10.1. Summary of physical, climatic and biological characteristics of southern islands, and of human impact upon them (data compiled mainly from: Holdgate & Wace, 1961; Abbott, 1974; Watson, 1975; Clark & Dingwall, 1985. Additional information from: Holdgate, 1965; Moore, 1968; Groves, 1981; Strange, 1981; Bell, 1982; Keage, 1982; Croxall et al., 1984b; Jouventin, Stahl, Weimerskirch & Mougin, 1984; Lewis Smith, 1984; Walton, 1985)

Island or island group	Structure and no. of main islands[a]	Total area (km²)	Maximum elevation (m)	Cover of permanent ice & snow (%)	Latitude (°S)	Mean annual temperature (°C)	Precipitation (mm)	Sector	No. of native vascular species (endemics)	No. of breeding birds (endemics)	No. of established introduced vascular plant species	No. of established introduced mammal species	Habitation (excluding sealing camps)[b]	Sovereignty
South Sandwich	G 11	618	1375	93	56–59	−1	500	Atlantic	1	15	0	0	ST	UK
South Georgia	SO	3755	2965	60	54	2	1500	Atlantic	26(1)	30(2)	22	3	(W:60); S	UK
Falklands	G 2	12030	760	0	51	6	635	Atlantic	163(14)	63(17)	91	11	F:1764	UK
Tristan da Cunha	G 3	115	2060	0	37	15	1700	Atlantic	93(18)	24(4)	115	7	F:1810	UK
Gough	S	65	910	0	40	11	3400	Atlantic	62(7)	23(2)	17	1	S	UK
Bouvetøya	SO	50	780	92	54	−1	500	Atlantic	0	8+	0	0	ST	Norway
Prince Edward	G 2	340	1230	0	46	5	2500	Indian	25	26	14	2	S	South Africa
Crozet	G 5	600	934	1	46	5	2400	Indian	28	36	45	4	S	France
Kerguelen	G 1	7200	1960	9	49	4	1100	Indian	30(1)	31	10	7	(W:24); S	France
St Paul	SO	7	272	0	38	13	1100	Indian	15	8	?	3	(F:5); ST	France
Amsterdam	S	55	910	0	37	13	1100	Indian	21+	9(1)	41	4	(F:1); S	France
Heard & McDonald	G 2	463	2745	82	53	1	1400	Indian	8	19	0	0	ST	Australia
Macquarie	S	118	433	0	54	4	900	Pacific	39(1)	21	4	4	S	Australia
Snares	G 3	3	152	0	48	11	1200	Pacific	18(1)	23	2	0	ST	New Zealand
Auckland	G 2	566	667	0	50	8	1500–2000	Pacific	187	46(1)	41	6	(F:14); ST	New Zealand
Campbell	SO	113	567	0	52	6	1450	Pacific	137	29(1)	81	3	(F:36); ST	New Zealand
Antipodes	G 1	21	402	0	49	8	1000–1500	Pacific	63(4)	25(1)	1	1	ST	New Zealand
Bounty	G 8	2	88	0	47	10	1000–1500	Pacific	0	7+(1)	0	0	ST	New Zealand

[a] G = group; SO = single island with offshore islets; S = single island.
[b] F = farming and date of establishment; W = whaling; ( ) = former activity and duration in years; S = permanent scientific station; ST = occasional scientific expeditions.

island with small offshore islets or a single island. The climate of each island varies according to the temperature of the sea surrounding it. The cool temperate islands lie between the Antarctic Convergence and the Subtropical Convergence. They are warmer and lack permanent ice and snow. The subantarctic islands lie within or on the Antarctic Convergence, and are colder and partially covered with glaciers and icecaps. Thus heavily glaciated islands like Heard and South Georgia are 'biological archipelagos' because their ice-free areas are separated by glaciers.

The number of vascular plant species found on each island is determined primarily by its temperature and secondarily by its area (Abbott, 1974). However, the plant communities of southern islands have a similar structure. The northernmost (Tristan, Gough, Amsterdam, Auckland and Campbell islands) have a scrub community, which is always of prostrate evergreen trees with an understorey rich in ferns. Most of the more southerly islands are dominated by a tussock grassland community and also have herbfield and fellfield communities. However, the floras on the coldest islands comprise lichens and mosses only. The distribution of vascular species found on each island reflects its position in the Atlantic, Indian and Pacific sectors of the Southern Ocean. None of the plant communities of southern islands has been grazed by terrestrial mammals during their development and therefore none are adapted to herbivory by vertebrates.

The faunas of southern islands are dominated by marine vertebrates. All islands abound in spectacular colonies composed of few species, but large numbers, of seals and seabirds. The seabirds are all ground- or burrow-nesting species and, apart from sometimes nesting on cliffs, lack defences against terrestrial vertebrate predators. In contrast, the terrestrial communities of southern islands contain an impoverished and disharmonic fauna. Communities are simple in terms of component species. The dominant macroscopic land fauna is composed of arthropods, either insects (principally springtails or Collembola) and arachnids (mainly mites and Acari) (Block, 1984). A few species of land-birds, mainly passerines and ducks, are found on some islands. Apart from the Falkland fox or 'warrah' (which is now extinct and was probably a feral dog: Clutton-Brock, 1977), southern islands entirely lack terrestrial mammals.

In spite of their relative poverty in terms of numbers of species, the floras and faunas of southern islands are of great biological importance (Holdgate & Wace, 1961). It is hard to define precisely the degree of endemism amongst various elements of the flora and fauna because

Figure 10.1. Mammals have been introduced to many southern islands. Herbivores, such as the rabbits introduced to Iles Crozet (A), provided fresh meat. Carnivores, such as the cats introduced to Macquarie Island (B), were companion animals. (Photographs by G. R. Copson and P. Vernon.)

taxonomic views change and research effort has been spread unevenly. This is especially so for arthropods, amongst which endemic species are considered to be common, but which are excluded from Table 10.1: relevant reviews include Gressitt (1970) and Block (1984). There are better estimates of numbers of endemic species amongst vascular plants and the avifauna of southern islands (Table 10.1). These are conservative estimates that include only those species, rather than the greater number of sub-species, currently thought to be endemic to particular islands or island groups. In addition, southern islands display an even greater degree of collective endemism because many of the species occurring on them are restricted in distribution to Insulantarctica.

It was for exploitation of marine mammals during the 18th century that man was first attracted to the southern islands. Gangs of sealers, who devastated fur seal stocks on all southern islands, usually occupied temporary camps on shore during southern summers (Bonner, 1982). Introductions of mammals to southern islands (Figure 10.1) began in the sealing era: accidental introductions of rodents from stores and ship-wrecks; deliberate introductions of herbivores to provide fresh meat and as insurance against being marooned. Farming enterprises were established on six island groups and a further range of species introduced. Exploitation of southern islands increased this century with establishment of shore-based whaling stations. The wheel of development has since turned full circle and now only the Falkland and Tristan da Cuhna islands are occupied permanently by established human communities. The remaining island groups are now occupied only by scientists. However, many of the introduced mammals remain as a legacy on southern islands and their effects on native floras and faunas have only recently begun to be appreciated and documented (Holdgate & Wace, 1961).

10.22   Introduced mammals

Mammals have been introduced deliberately to 15 islands or island groups (Figures 10.1, 10.2). Introduced populations resulting from either accidental or deliberate introductions are presently established on 13 islands or island groups (Table 10.2). Species introduced accidentally comprise three rodents: black rats, Norway rats and house mice. Species introduced deliberately include several of man's domesticated and husbanded animals. They provided farm stock, familiar sources of fresh meat, beasts of burden, companion animals and control of rodent pests. Lagomorphs (usually rabbits, and hares on only one island group) were

Table 10.2. *Date of introduction, present extent of distribution and status of introduced species established on southern islands. L = localised; W = widespread; C19 = 19th century; * = husbanded; † = feral (shown only for Falkland Islands and Tristan da Cunha); [1963] = date of count (data compiled mainly from Holdgate & Wace, 1961; Clark & Dingwall, 1985)*

Island or island group	Accidental Introductions			Deliberate Introductions														Additional sources
	Black rats	Norway rats	House mice	Cattle	Horses	Donkeys	Reindeer	Guanaco	Sheep	Mouflon	Goats	Rabbits	Hares	Pigs	Dogs	Cats	Silver foxes	
South Georgia		c. 1800 W: many areas	C19 L: 1 area				1911 & 1926 L: 3 areas 2000 [1976]											Pye & Bonner (1980); this study
Falklands	C18 Widespread	C18 Widespread	Widespread	1764† Widespread 8200 [1979]	1764* Widespread 2300 [1979]			c. 1930* Localised 60 [1970]	1852* Widespread 6×10⁵ [1975]			1764† Widespread Localised	?†		?* Localised	C18*† Widespread	c. 1930† Localised	Strange (1981)
Tristan da Cunha	1882 L: Tristan		C18 L: Tristan	c. 1820*† L: Tristan 330 [1968]		c. 1860* L: Tristan			c. 1820*† L: Tristan 260 [1968]						c. 1820* L: Tristan	c. 1810*† L: Tristan		Wace & Holdgate (1976)
Gough			c. 1880 Widespread															Wace & Holdgate (1976)
Prince Edward			C19 W: Marion													1949 W: Marion 600 [1982]		van Aarde (1979)
Crozet	C19 L: Possession		C19 L: 2 islands									c. 1850 L: Cochons 2000 [1975]				1887 L: Cochons 100 [1975]		Derenne & Mougin (1976)
Kerguelen	c. 1850 L: Grande Terre		C19 W: Grande Terre				1955 W: 2 islands 2000 [1974]		1909 L: 3 islands 800 [1975]	1957 L: Haute 300 [1977]		1874 W: 4 islands 177000 [1968]				1951 W: Grande Terre 3500 [1975]		Pascal (1982)
St Paul	C18 Widespread		C19 Widespread	C19 Widespread								C19 Widespread						Segonzac (1972)
Amsterdam	C18 Widespread		C19 Widespread	1871 Widespread 2000 [1964]												C19 Widespread		Dorst & Milon (1964); Leset (1969)
Macquarie	C18 Widespread		C19 Widespread	C19 Widespread								1879 Widespread 150000 [1972]				c. 1820 Widespread 250–500 [1975]		Jones (1977); Jenkin et al. (1982)
Auckland			C19 W: 2 islands	1895 L: Enderby 50–60 [1983]							1850 L: Auckland 100 [1983]	1840 L: 2 islands 2000 [1983]		1807 L: Auckland		C19 L: Auckland		Taylor (1971); Rudge & Campbell (1977); Challies (1975); Dingwall (1988)
Campbell		C19 Widespread							1895 Localised 800 [1984]							c. 1920 Widespread Scarce [1977]		Dilks (1979); Dingwall (1988)
Antipodes			?C19															

introduced widely to provide fresh meat. More exotic ungulates, namely reindeer, mouflon and guanaco, have been introduced to one or two islands. Fur-bearing predators have also been introduced occasionally. The species involved were mink (Falkland Islands and Iles Kerguelen), otters and skunks (Falkland Islands) and silver foxes (South Georgia and Falkland Islands).

Figure 10.2. Outcome of attempts to introduce mammals to southern islands, shown both for different islands or island groups and for different species. (Data compiled mainly from Holdgate & Wace, 1961; Clark & Dingwall, 1985. Additional information from: Dorst & Milon, 1964; Holdgate, 1965; Taylor, 1971; Segonozac, 1972; Challies, 1975; Lesel & Derenne, 1975; Wace & Holdgate, 1976; Rudge & Campbell, 1977; Strange, 1981; Jenkin *et al.*, 1982; Keage, 1982; Pascal, 1982; Headland, 1984).

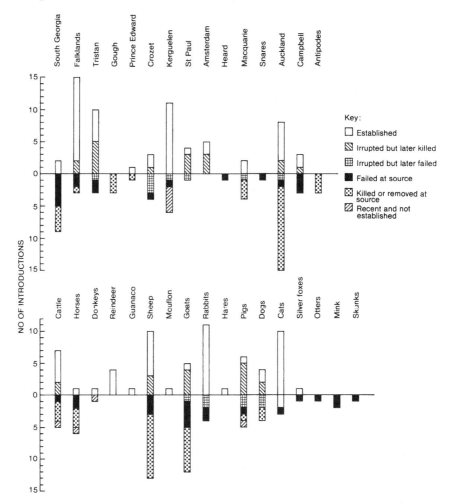

Deliberate introductions to the southern islands have been made on at least 121 occasions. An assessment of why not all the introductions are now established on the islands (Figure 10.2) suggests that mammals from *c.* 25% of introductions were killed or removed without ever being given a reasonable chance to establish feral populations. The majority in this category were those introductions kept housed or around settlements, needed quickly for meat (mainly domestic stock) or which proved uneconomic to maintain (mink on the Falkland Islands or foxes on South Georgia). Very few in this category were killed because it was realised that the introductions would be detrimental to the islands' ecology (introductions to Gough, Marion in the Prince Edward group and Macquarie islands provide the few examples). Even now, domestic animals (cattle, pigs, horses and donkeys) still live around the scientific station at Iles Kerguelen, but have not established themselves.

Introductions to southern islands appeared to fail at two different stages. About 13% of attempts failed soon after introduction. Included amongst these were most introductions to South Georgia, sheep on Heard Island and most introductions of fur-bearing animals (Figure 10.1). In contrast, species arising from *c.* 8% of introductions were at one time common but subsequently died out by themselves. Included amongst these were goats and pigs on Inaccessible Island in the Tristan group and cats on Ile de la Possession in the Crozet group and on Ile St Paul. The last category of failed introductions have been differentiated from the *c.* 12% of introductions in which the species was also at one time common, but in which hunting played a major role in their demise. Were it not for human interference, these introductions would probably be amongst those still established on southern islands.

The species that have been introduced (Figures 10.1, 10.2) have a broad spectrum of requirements for their survival in terms of temperature tolerance and food preference. In contrast to introductions of reindeer made to Arctic islands (Chapter 8), several species were taken outside their normal climatic range when introduced to southern islands. Without both knowledge of the sex ratio and reproductive status of the animals introduced and detailed study of the fate of each introduction it is perhaps hazardous to generalise about reasons underlying their failure or success. However, temperature is the most important determinant of the biological characteristics of southern islands (Abbott, 1974). Interestingly, island temperature also correlates with the success ratio of introductions to different islands (Figure 10.3). Thus, the rigorous temperature of southern islands, and the ecological characteristics which themselves correlate with temperature, have probably played an important role in limiting the success of introductions. In relative terms, the warmer the island the more successful have been attempts to establish introduced species.

The islands or island groups upon which introductions are now established fall into two distinct categories, dependent upon their structure (Table 10.1). The first includes those where a single introduction has become widespread across the whole group because there is only one main island and no physical barriers to prevent dispersion (Gough, St Paul, Amsterdam, Macquarie and Campbell islands). The second category is those which are either paired or a group or dissected into 'islands' by large glaciers. In this category single introductions have resulted only in localised distributions (Table 10.2). Thus Inaccessible and Nightingale islands in the Tristan group, Prince Edward Island, three islands in the Crozet group and Adams Island in the Auckland group remain free of introduced mammals. Rats on South Georgia, introductions on the Falkland Islands and rabbits on Iles Kerguelen are widespread only as a result of multiple introductions to different islands or ice-free areas.

Accounts of the outcomes of deliberate introductions of mammals to southern islands were interpreted and classified into six categories (Figure 10.2). Phrases like 'were at one time common' were taken to mean that an irruption had taken place, and so on. Such interpretations

Figure 10.3. Correlation between mean annual temperature and the success ratio of introductions on southern islands (data from Table 10.1 and Figure 10.2).

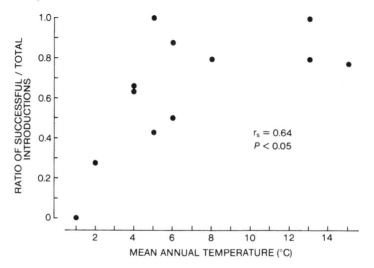

$r_s = 0.64$
$P < 0.05$

are not infallible, but unlikely to be far wrong (cf. Chapter 8). Introductions to different islands or areas in the same island group were each scored in the analysis of total introductions attempted. However, only one introduction (the most successful) to one island or area within the group was scored in this analysis. Introductions that were killed or removed at source were excluded from the total of failed attempts in the calculation of the ratio of successful to total introductions. A Spearman rank correlation coefficient, corrected for ties, shows the relationship between this success ratio and island temperature is significant ($r_s = 0.64$, $t = 2.63$, $df = 10$, $P < 0.05$).

10.3 Establishment of an introduction

Assuming that both climatic conditions and the range of food on an island are suitable for its requirements, an introduced mammal generally faces an ideal situation in which food is plentiful, competition from native species is lacking and predators are absent. As suggested above, most introductions were made too long ago for their early history to be certain. However, studies have been carried out on a number of islands in recent years. In this section I review some of the important aspects of the biology of introduced mammals that have led to their establishment on southern islands.

10.31 Breeding

Several of the species introduced to southern islands were already adapted to seasonal environments and had distinct breeding seasons in their areas of origin. Amongst these species are reindeer, sheep and mouflon, and cattle. However, those introductions made directly from the Northern Hemisphere had to reverse their breeding seasons to produce young in the austral spring (Figure 10.4). Several of the other introduced species have a less distinct, or no, breeding season in their native habitats and can give birth all year round if conditions are favourable. On the colder subantarctic islands, from which most data are available, nearly all introduced species have developed distinct breeding seasons (Figure 10.4). For cats and rodents this still allows sufficient time for more than one litter to be born per year (e.g. van Aarde, 1978; Pye & Bonner, 1980; Berry *et al.*, 1979; Pascal, 1982; Pye, 1984). It does, however, still represent a considerable restriction compared with potential breeding seasons in more equable climates. The restriction probably arises from the pronounced changes in photoperiod occurring at relatively low latitudes, and especially from marked seasonal changes in the availability of food and, therefore, in body condition. Comparative studies of factors determining timing of the breeding season of introduced mammals on southern islands would clearly be of great interest.

10.32 Genetic adaptation

Each population of introduced mammals on southern islands is probably genetically unique. Each has tended to be founded by a small group of animals, and most populations have remained isolated since their introduction. Mice are the best studied species on southern islands. The amount of brown fat found increases markedly on the coldest islands (Berry *et al.*, 1979). There is strong presumptive evidence that cold stress acts as an important selection pressure, and gene loci that varied between mice in different age groups and on different islands have been identified (Berry *et al.*, 1978, 1979). Cats also show strong founder effects (Dreux, 1974; van Aarde & Robinson, 1980). For example, the frequency of alleles conferring a dark phenotype is higher on Marion Island than in South Africa where the cats originated. Some adaptive advantage may be conferred upon mutant phenotypes, which increase in frequency with age (van Aarde & Skinner, 1982). However, it still remains for genetic findings to be related to ecological adaptations of mice and cats on southern islands.

The genetics of the ungulate and lagomorph populations remain almost entirely unstudied. There are suggestions of founder effects amongst coat colours of rabbits on Iles Kerguelen, and of genetic adaptations in fleece properties amongst sheep on Campbell Island

Figure 10.4. Months when births occur for introduced mammals on subantarctic islands. (Data from: this study; Berry, Bonner & Peters, 1979; Pye & Bonner, 1980; Berry, Peters & van Aarde, 1978; van Aarde, 1979; Derenne & Mougin, 1976; Lesel & Derenne, 1975; Derenne, 1976; Jones, 1977; Skira, 1978; Pye, 1984.)

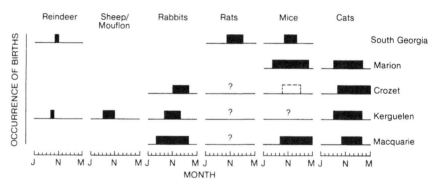

(Anon., 1983; Bousses, Chapuis & Arthur, 1988). There is also a possibility that the high prevalence of dental abnormalities amongst reindeer on South Georgia might be associated with inbreeding (Chapter 7). Some introductions may also have failed at source due to inbreeding depression and high juvenile mortality (cf. Ralls *et al.*, 1979). Thus, introduced mammals on southern islands offer interesting and largely untapped possibilities for studies that relate genetics to adaptation and ecology.

10.33 Population growth

The small number of animals introduced originally is also of importance to subsequent patterns of population growth. Unfortunately, many of the introductions were made before reliable census methods were devised. In any case, small species like rodents, lagomorphs and cats are hard to count. Hence, reliable data describing changes in population size of introduced mammals on southern islands are limited. Of the introductions made to southern islands before the 1950s, the only good data are from counts of sheep on Campbell Island (Figure 10.5). There are also estimates of increases in population size of the more recent introductions to Iles Kerguelen (Lesel & Derenne, 1975; Pascal, 1982), and of cats on Marion Island (Figure 10.5).

These data, and the many more anecdotal accounts of changes in population size of introduced mammals on southern islands, suggest that the irruptive model described in Chapters 2 and 8 may be appropriate for

Figure 10.5. Changes in population size of sheep on Campbell Island and cats on Marion Island. Dots represent counts, the solid line shows the likely change in population size, and the broken line the change in population size after management procedures (erection of a fence and introduction of feline panleucopaenia virus) had been adopted (data from Wilson & Orwin, 1964; Clark & Dingwall, 1985; van Aarde, 1979).

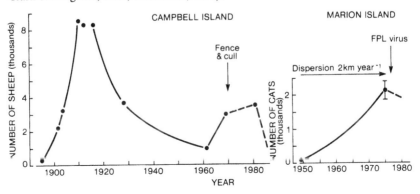

most ungulate and non-ungulate species. Suggestions to this effect have been made for pigs on Auckland Islands and cats on Iles Crozet (Challies, 1975; Derenne & Mougin, 1976), and for rabbits on other islands (e.g. Schlanger & Gillett, 1976).

The irruptive model helps in the consideration of many aspects of the ecology and management requirements of introduced mammals on southern islands. After introduction numbers will initially show a rapid increase whilst food is abundant. Later, numbers will reach an irruptive peak and then decline to a more stable level. This occurs because introduced mammals reduce, and then reach a balance with, their food supply. Therefore, it may be inferred that recent introductions (reindeer, mouflon and cats on Kerguelen, or cats on Marion) have yet to reach peak numbers. If management is attempted at this stage, the goal is to contain the irruption (Caughley, 1981). In contrast, the well-established introductions are likely to show only the more dampened fluctuations characteristic of natural populations. Numbers reached at the peak of the irruption will probably not be attained again since the initially abundant food supply is unlikely to be fully restored in the continued presence of the introduced mammal. Thus, reindeer from the Barff and Busen herds on South Georgia have declined and reduced the cover of their preferred forage species (Figure 8.8, Chapter 9). Sheep on Campbell Island and rabbits on the Auckland Islands have also declined from peak numbers for similar reasons (Figure 10.5; Taylor, 1971). Therefore, population numbers derived from recent censuses of long-established mammals are unlikely to change dramatically (Table 10.2), unless there is outside interference from man (such as the erection of a fence on Campbell Island). If management is attempted at this stage, the goal will be to alter a displaced equilibrium to the advantage of native species and the disadvantage of introduced species (cf. Caughley, 1981).

10.34   *Feeding*

The species introduced to southern islands comprise herbivores, omnivores and carnivores. This, coupled with their wide range of body sizes, suggests a broad spectrum of food requirements amongst the various species. However, the foods eaten by many of the mammals on southern islands have not been studied directly. Instead, there is a larger body of comment on the effects of the mammals on native floras and faunas (discussed more fully in 10.4).

The available data on foods eaten by introduced mammals again derive mainly from the colder subantarctic islands (Table 10.3). The grazing

Table 10.3. Examples of the main and preferred foods of introduced mammals ... quantitative data are available, items marked * predominate in the diet

| Island | Introduced animal | | | | | | | | | Source |
	Cattle	Reindeer	Sheep/mouflon	Goats	Rabbits	Pigs	Rats	Mice	Cats	
South Georgia		Poa flabellata* Acaena magellanica Lichens					Poa flabellata* Invertebrates Carrion	Poa flabellata* Invertebrates		This study Pye & Bonner (1980); Berry et al., (1979)
Marion								Invertebrates* Acaena magellanica Poa cookii	Petrels* Vegetation Mice	Gleeson & van Rensburg (1982); van Aarde (1980)
Crozet					Acaena magellanica Pringlea antiscorbutica		?	Plants* Carrion	Prions & petrels* Mice Penguins	Derenne & Mougin (1976)
Kerguelen		Acaena magellanica Azorella selago Acaena Pringlea antiscorbutica Lichens	Pringlea antiscorbutica Acaena magellanica		Poa cookii Azorella selago Pringlea antiscorbutica		?	Azorella selago Acaena magellanica	Prions & petrels* Rabbits Vegetation	Lesel & Derenne (1975); Pascal (1982); Derenne (1976)
Macquarie					Poa foliosa Stilbocarpa polaris		Poa foliosa Invertebrates Carrion	Stilbocarpa polaris Poa foliosa Invertebrates	Rabbits* Prions & petrels	Costin & Moore (1960); Jenkin et al., (1982); Pye (1984); Jones (1977)
Auckland	Pseudopanax simplex Coprosma foetidissima Myrsine divaricata			Pseudopanax simplex* Coprosma foetodissima Metrosideros umbellata	Poa annua* Scirpus aucklandicus* and others	Pleurophyllum hookeri Seaweed Invertebrates Birds		?		Taylor (1971); Rudge & Campbell (1977); Challies (1975)
Campbell			Chionochloa antarctica Poa litorosa Pleurophyllum spp. Stilbocarpa polaris				?		Rats* Insects Introduced birds Native birds	Wilson & Orwin (1964); Dilks (1979)

herbivores prefer the various species of tussock grass and broad-leaved herbs, where these are still available. Indirect evidence and one study of cattle on Amsterdam suggest that cattle and goats browse heavily upon the woody species found in scrub and rata forest communities on the warmer southern islands (Holdgate, 1965; Lesel, 1969; Taylor, 1971; Rudge & Campbell, 1977). Rodents are omnivorous (Table 10.3). Mice eat both vegetable matter, especially seeds, and invertebrates. Rats also prey upon native birds, mostly small ground-nesting passerines and burrow-nesting petrels, but some of their food may be carrion rather than prey. Pigs are also omnivores and both graze upon broad-leaved herbs and root for burrow-nesting birds (Table 10.3). Cats are the only carnivore whose diet has been studied on southern islands. Their prey depends to some extent upon whether rabbits and rodents have been introduced to the same island. Burrow-nesting petrels and prions, passerines and ducks are the main native species preyed upon.

As with changes in population size, the irruptive model provides a useful framework in which to set studies of feeding of introduced mammals on southern islands. Upon introduction, food is plentiful and introduced mammals can select preferred (or for carnivores most easily caught) species. As preferred species become reduced in density and availability, introduced mammals have to adjust their diet to less palatable (or less easily caught) species. This adjustment has been shown already for reindeer on South Georgia (Chapter 8), and similar changes in diet have been discussed for other herbivores and an omnivore on

Figure 10.6. Diet of cats on various southern islands (shown with numbers of years since their introduction). Pairwise comparisons between recent and long-established introductions have been made only for islands having either rabbits or mice as the dominant introduced prey (data from Derenne, 1976; Derenne & Mougin, 1976; Jones, 1977; Dilks, 1979; van Aarde, 1980).

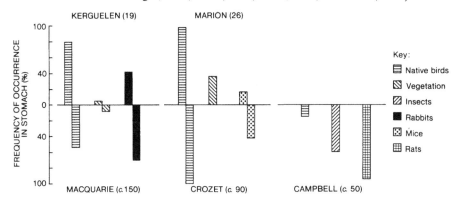

southern islands (Wilson & Orwin, 1964; Challies, 1975). Comparisons between recent and well-established introductions suggest that cats also make similar adjustments. Their diets show distinct prey shifts from easily caught native birds to less easily caught introduced mammals (Figure 10.6). This pattern has been confirmed further by changes observed in the diet of cats in studies separated by a few years on one island (Pascal, 1980; van Rensburg, 1985). Therefore, species upon which introduced mammals are said to have had an impact may no longer be found in present day dietary studies (see Burger, 1985).

By chance man has helped the dietary adjustment of many introduced herbivores and carnivores. Plants have been introduced to most southern islands by accident (Table 10.1). Grazing-tolerant species like *Poa annua* are now widespread in plant communities on all islands associated with long-established introductions (Walton, 1975). In turn these species are now amongst the preferred forage of many introduced herbivores (Figure 10.7A). Similarly, the presence of rabbits and rodents as alternative prey has probably sustained many of the long-established populations of cats on southern islands (Figure 10.6), as well as populations of native avian predators (Figure 10.7B).

10.4   Effects on native biotas

The floras and faunas of islands with simple and species-poor communities are particularly prone to outside causes of disturbance. The southern islands are no exception and their biotas have altered dramatically since the arrival of man. In this section, I review some of the consequences arising from the introduction of mammals to southern islands.

10.41   Floras

The grazing of herbivores has led to the local reduction of the most palatable species amongst the native floras. Tussock grasslands and herbfields have generally been the most disturbed communities and individual species have been heavily browsed or grazed in other habitats (Table 10.4A). All changes observed amongst native species comprise a reduction in density and cover of plants preferred by introduced herbivores, and the spread of unpalatable species like *Bulbinella rossii* on Campbell and Auckland islands. There is no firm evidence to suggest that any native species has been totally eliminated from southern islands occupied by introduced herbivores. The only possible exception is *Poa flabellata* which was originally suggested to have been eliminated from

Figure 10.7. Introduced species form alternative prey on southern islands. The grazing tolerant grass *Poa annua* now forms grazing lawns on all islands with long established introductions of herbivores such as Macquarie Island (A). Rabbits themselves fall prey to introduced cats and to native predators like the great skua (B). (Photographs by G.R. Copson.)

Tristan Island (Holdgate & Wace, 1961). However, later work suggested it was never present, even though it is abundant on Gough Island (Wace & Holdgate, 1976; Groves, 1981). Hence, data from other southern islands generally support conclusions from New Zealand and South Georgia that extinctions of native species do not occur through foraging by introduced herbivores (Caughley, 1981; Chapter 9).

Another finding common to all southern islands is the extensive incorporation of grazing-tolerant introduced species into native plant communities. *Poa annua* is found universally on all southern islands with heavily grazed communities (Figure 10.7A). Other species are also found in grazing lawns on various southern islands. Colonisation of native communities by introduced species stabilises the soil and prevents erosion. The actual causes of erosion on southern islands are presently being debated (Chapter 9). Early work suggested that herbivores were an important cause (Holdgate & Wace, 1961). However, studies on Macquarie Island now question this conclusion (Griffen, 1980; Scott, 1983; Selkirk *et al.*, 1983). Comparative studies of the cause of erosion on other southern islands would be well merited.

10.42 Faunas

The faunas of southern islands are affected by the various types of introduced mammals in different ways. Reduction in cover of tussock grassland caused by herbivores results in loss of nesting sites for many species of seabirds. This is especially obvious on the Falkland Islands where, due to that and to the activities of man, seabird colonies are now largely restricted to offshore islets (Table 10.4B). Predation of birds by omnivores and carnivores has also been of major importance (Johnstone, 1985). With such different combinations of introduced mammals present on the various islands, it is difficult to disentangle the differing impacts of each species, except that mice have been of little importance (Figure 10.8). The species amongst native avifaunas that have been most affected are land birds and burrow-nesting prions and petrels (Table 10.4B). These species are small enough to be caught by introduced predators such as cats and rats (Moors & Atkinson, 1984; Atkinson, 1985). Pigs also dig up burrows with ease (Challies, 1975). Larger seabirds like penguins and albatrosses less frequently fall prey to introduced mammals.

In terms of biomass, the greatest loss amongst avifaunas has been that of prions and other petrels. Studies on Marion Island, in particular, have shown that large numbers of birds are eaten by cats (van Aarde, 1980).

Table 10.4. Examples of ways that introduced mammals have affected the floras and faunas of southern islands

A. Effect of introduced herbivores and omnivores on floras

Affected species, group or community	Status	Eliminated or extinct	Locally reduced	Locally spreading	Island or island group	Animal responsible	Reason	Source
Scrub								
Phylica arborea	Native		+		Tristan, Amsterdam	Cattle	Browsing	Holdgate & Wace (1961); Lesel (1969)
Rata forest								
Metrosideros umbellata *Myrsine divaricata* *Pseudopanax simplex*	Native		+		Auckland	Cattle, goats	Browsing	Taylor (1971); Rudge & Campbell (1977); Campbell & Rudge (1984)
Tussock grassland								
Poa flabellata	Native		+		Falklands, South Georgia	Cattle, sheep, Reindeer	Grazing	Holdgate & Wace (1961); Davies (1939); this study; Lesel & Derenne (1975); Costin & Moore (1960); Wilson & Orwin (1964); Dilks & Wilson (1979); Challies (1975); Campbell & Rudge (1984)
Spartina arundinacea			+		Tristan	Cattle	Grazing	
Poa cookii			+		Kerguelen	Rabbits	Grazing	
Poa foliosa			+		Macquarie, Auckland	Rabbits, goats	Grazing	
Chionochloa antarctica			+		Campbell	Sheep	Grazing	
Poa litorosa				+	Auckland	Pigs, goats	Unpalatable	
Heath								
Acaena magellanica	Native		+		South Georgia, Kerguelen	Reindeer	Grazing	This study; Lesel & Derenne (1975)
Acaena magellanica			+		Crozet	Rabbits	Grazing	
Azorella selago	Native		+	+	Kerguelen	Rabbits	Dispersion of seeds Grazing	Lesel & Derenne (1975)
Herbfield								
Arabis macloviana *Calandrinia feltonii*	Endemic		+		Falklands	Sheep	Grazing	Moore (1968)
Pringlea antiscorbutica	Native		+		Kerguelen, Crozet	Rabbits	Grazing	Lesel & Derenne (1975); Derenne & Mougin (1976); Costin & Moore (1960)
Stilbocarpa polaris *Pleurophyllum* spp. *Anistome latifolia*	Native		+		Macquarie, Auckland, Campbell	Rabbits, Pigs, goats, Sheep	Grazing	Challies (1975); Rudge & Campbell (1977); Wilson & Orwin (1964); Taylor (1971); Campbell & Rudge (1984)
Bulbinella rossii	Native			+	Auckland	Sheep, Cattle, rabbits	Unpalatable	
Macrolichens	Native		+		South Georgia, Kerguelen	Reindeer	Grazing	Lindsay (1973); Lesel & Derenne (1975)
Poa annua and others	Introduced			+	All islands	All herbivores	Grazing-tolerant	Walton (1975)

B. Effect of introduced mammals on faunas

Affected species, group or community	Status	Eliminated or extinct	Locally reduced	Locally spreading	Island or island group	Animal responsible	Reason	Source
Seabirds (most species)	Native		+		Falklands	Rabbits, sheep	Loss of nesting sites in tussock grass	Croxall, McInnes & Prince (1984a)
Burrow-nesting prions and petrels (most species)	Native		+		All islands	Cats, rats, pigs	Predation	Johnstone (1985)
Pipit *Anthus antarcticus*	Endemic		+		South Georgia	Rats	Predation	Pye & Bonner (1980)
A. novae-zelandiae	Native		+		Campbell	Cats, rats	Predation	Johnstone (1985)
Bunting *Nesospiza acunhae*	Endemic	1873						
Moorhen *Gallinula n. nesiotes*	Endemic	C19			Tristan	Cats, rats	Predation	Wace & Holdgate (1976)
Thrush *Nesocichla e. eremita*	Endemic		+					
Tern *Sterna virgata*	Native		+		Marion, Kerguelen	Cats	Predation	
Parakeet *Cyanoramphus novae-zelandiae*	Endemic	1880–91			Macquarie	Cats/wekas	Predation pressure after arrival of rabbits	Taylor (1979)
Rail *Rallus philippensis*	Endemic	1894						
R. pectoralis muelleri	Endemic		+					
Snipe *Coenocorypha a. aucklandica*	Endemic		+		Aucklands	Cats, rats	Predation	Johnstone (1985)
Merganser *Mergus australis*	Endemic	C19						
Teal *Anas aucklandica nesiotis*	Endemic		+		Campbell	Cats, rats	Predation	Johnstone (1985)
Skua *Catharacta skua*	Native			+	Macquarie, Kerguelen	Rabbits	Alternative prey	Lesel (1968); Jones & Skira (1979)
Weka *Gallirallus australis*	Introduced			+	Macquarie	Rabbits	Prey	Jenkins et al. (1982)
Cats	Introduced			+	Several islands	Rabbits, rats, mice	Alternative prey	see Figure 10.5

However, in terms of species, the most serious losses have been amongst the land birds (Table 10.4B). Several endemic species and sub-species have become extinct and others are found only on offshore islands free of introduced mammals. The loss of endemic birds on southern islands is in direct contrast to the lack of recorded loss amongst plants caused by herbivores. A particularly notable pair of extinctions was caused indirectly by a non-predatory introduced mammal. Banded rails and parakeets survived on Macquarie Island for the first 70 years after the arrival of cats, but declined to extinction after the introduction of rabbits. This latter introduction led to great increases in the number of feral cats and of introduced wekas, and in turn to more intensified predation on the endemic birds (Taylor, 1979).

Research on the effects of introduced mammals on the less noticeable elements of native faunas has lagged behind that on avifaunas. However, some quantitative data are now becoming available. For example, grazing and trampling of the soil by reindeer on South Georgia alters the composition of the arthropod fauna (Figure 10.9). Of particular note is the reduction in the perimylopid beetle, *Hydromedion sparsutum*, which is a primary decomposer (Vogel, Remmert & Lewis Smith, 1984). Grazing of *Pringlea antiscorbutica* by rabbits on Kerguelen has seriously reduced populations of invertebrates associated with it (Holdgate, 1967).

Figure 10.8. The effects of various combinations of introduced predators upon landbirds and burrow-nesting petrels and prions on southern islands (Data from: Croxall *et al.*, 1984a,b; Jouventin *et al.*, 1984; Rounsevell & Brothers, 1984; Williams, 1984; Tables 10.2 & 10.4B.)

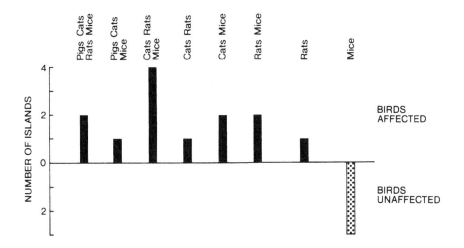

Further topics that might be addressed include the following: (1) what effect do rodents have upon populations of their invertebrate prey? (2) does the role of the native microfauna in nutrient cycling change in the presence of herbivore dung? (3) are native invertebrates affected by intermediate stages of parasites brought in by herbivores? This interesting and important field of study still awaits development on southern islands.

10.43   Reversibility of change

Introduced mammals have had an undoubted impact on the biotas of southern islands. This raises the important question of whether or not these changes are irreversible. Suggestions were often made to the effect that changes on islands were either permanent or likely to take many years to reverse (Elton, 1958; Holdgate & Wace, 1961). Obviously, extinctions of endemic species and sub-species amongst native avifaunas are permanent. However, historical records and some recent experiments carried out on southern islands suggest that a more optimistic outlook can be adopted for the reversibility of change amongst other elements of the native biotas.

Large scale experiments that remove grazing pressure have been conducted on Campbell and Macquarie islands. A fence was erected across Campbell Island in 1970 and sheep to the north of it were destroyed. This resulted in a dramatic change in the vegetation, and regeneration of preferred native plants has been vigorous (Dilks & Wilson, 1979; Figures 10.10, 10.11). The introduction of myxomatosis to Macquarie in 1978 has reduced the rabbit population substantially, and

Figure 10.9. Differences in the composition of the invertebrate fauna of South Georgia in areas ungrazed by (U), exclosed from (E), and still grazed by (G), reindeer (data from Vogel *et al.*, 1984).

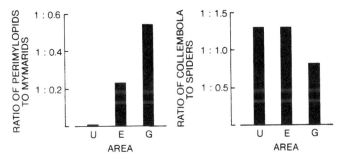

Figure 10.10. The vegetation of Campbell Island has recovered following the erection of a fence in 1970 and exclusion of sheep from the north side of the island. Grazing by sheep resulted in the reduction of native tussock grass, the formation of short turf and an increase in the unpalatable *Bulbinella rossii* (foreground); native tussock grass, *Poa litorosa*, has regained dominance on the far side of the fence (A). Megaherbs, notably *Pleurophyllum speciosum*, are now common amongst the tussock grass (B). (Photographs by C.D. Meurk.)

native vegetation is beginning to recover at the expense of introduced species (Brothers, Copson, Skira & Eberhard, 1982; Scott, 1983). Results from these islands and from the exclusion experiment on South Georgia (Chapter 9) suggest that vegetation on southern islands can, for the most part, recover to an approximation of its former state within a decade of grazing pressure being removed. These three experiments are supported by many more anecdotal accounts of the recovery of vegetation on various islands or island groups when herbivores are removed (e.g. Headland, 1984). It is also of note that some elements of the arthropod fauna can revert to their former composition in reindeer exclosures (Figure 10.9).

Unfortunately there is no comparable experimental evidence to date that shows whether native avifaunas can recover their former abundance on southern islands if predators are eradicated. Results may be expected shortly from Marion Island, where a programme to eradicate cats by a combination of introducing feline panleucopaenia virus and shooting is under way (van Aarde & Skinner, 1982). Cat eradication programmes on islands to the north of New Zealand have resulted in spectacular recoveries in numbers of affected seabirds (Moors & Atkinson, 1984;

Figure 10.11. Changes amongst palatable herbs and tussock grass on Campbell Island since the construction of a fence in 1970, and exclusion of sheep from the north side of the island. Numbers of native plants were counted in permanent belt transects laid out to the north and south of the fence in 1975 and 1980 (data from Meurk, 1982).

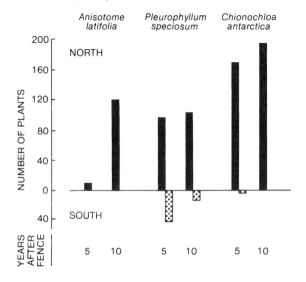

Veitch, 1985). Apart from extinctions, this evidence suggests that an optimistic outlook is appropriate for the recovery of native avifaunas on southern islands if predators are removed successfully.

In conclusion, introduced mammals on southern islands are of great potential interest for scientific study, and much still needs to be learned of their effects on indigenous floras and faunas. Recent evidence suggests that southern island ecosystems generally have great powers of recovery if the influence of introduced species is removed.

10.5   Summary

1. Southern islands are remote and cold. They have impoverished native terrestrial floras and faunas, but substantial populations of seabirds. Their floras and avifaunas lack defences against herbivory and predation by terrestrial vertebrates.

2. The islands were all exploited for seal hunting by man from the 18th century and onwards. Mammals were introduced both accidentally and deliberately from then until the 1950s. Island temperature has been important in determining whether introductions fail or succeed.

3. Introduced mammals appear to have become established on southern islands during the course of an irruptive oscillation. A total of 17 species, ranging in size from rodents to cattle, are now established on 13 islands or island groups.

4. The species introduced comprise herbivores, omnivores and carnivores. Herbivores have reduced the cover of native plants and facilitated the spread of introduced grazing-tolerant species, but not caused any extinctions. In contrast, omnivores and carnivores have both reduced burrow-nesting seabird colonies and caused extinctions of endemic land birds.

5. Apart from extinctions, many of the changes caused to native biotas by introduced mammals are likely to be reversible if introduced mammals are removed from southern islands.

Part IV

Overview

Reindeer and caribou are a single species of deer with an holarctic distribution. They are well adapted to life at high latitudes. In their Arctic habitats, the available forage comprises a diverse range of plant species and communities that are adapted to grazing by large herbivores. Lichens form a major part of the winter diet of continental populations. Although overgrazing can, and has, occurred in localised areas, annual migrations and low overall densities of reindeer and caribou generally ensure that lichens are not overutilised.

Reindeer and caribou are of great economic and cultural importance to northern peoples. Man has herded reindeer in Eurasia and hunted caribou in North America for many centuries and their annual migrations have established a nomadic pattern of life for Lapps and other ethnic groups. The domestication of reindeer has reduced both the number and size of wild populations which now have a more restricted range throughout Eurasia.

The importance of reindeer and caribou to man has resulted in the initiation of a wide range of studies in the Arctic. However, the direct influence of man upon most herds has sometimes made it difficult to investigate the ecology of the species in the wild. Reindeer also have been introduced to a number of different islands. This has given rise to an unusual variety of natural experiments in which it has been possible to study population dynamics in situations uninfluenced by man and to investigate the vulnerability of island ecosystems to introduced herbivores.

This study focuses on the Norwegian reindeer that were introduced to the subantarctic island of South Georgia in 1911 and 1925 (Endpiece). There were no predators or competitors and an abundant, but unusual, food supply. The impoverished flora had a low standing crop of lichens and plant communities were not adapted to grazing by large herbivores.

The reindeer responded by undergoing an irruption, which led in due course to overgrazing of native plant communities.

South Georgia's vegetation is dominated by the coastal tussock grass *Poa flabellata*, a species of high biomass and productivity and with leaves that contain a high content of soluble carbohydrate. Although other plant communities are available for summer grazing, only tussock grassland is accessible through snow in winter. Not only did the reindeer adapt in changing their diet to a new range of species, but they rapidly came to depend entirely on *P. flabellata* rather than lichens for winter forage. Reindeer also reversed their breeding season by six months on introduction to South Georgia, in order to calve in the austral spring.

Reindeer and caribou have a polygynous social organisation, live in large groups and females are unique amongst cervids in having antlers. The species is also amongst the more sexually dimorphic of contemporary cervids, and this often leads to sex differences in ecology and life history. On South Georgia the quality of forage eaten by males and females showed few differences. However, male reindeer achieve physiological puberty as calves, whilst females only achieve puberty as yearlings when over 90% became pregnant. The antler cycles of males and females differ in their timing due to the influence of testosterone in males, but a non-gonadal hormone common to both sexes may exert the principal control over the cycle. Males also produce much heavier and more complex antlers than females relative to their body weight, and males lose twice as much of their body weight during the annual cycle as females. This sexual dimorphism in body weight and in the costs of reproduction results in a differential mortality between the sexes.

The mortality of reindeer on South Georgia follows the general pattern for mammals. Calf mortality is significant, mortality rates then decline but increase again in old age. As this population is food-limited, most deaths occur in periods that coincide with minimum body weights in each sex, in early winter after the rut for males and in late winter around parturition for females. Many reindeer fall to their death over cliffs because the accessibility of tussock grass in winter is now becoming increasingly restricted to steep slopes and cliffs. Indeed, a sequence of overgrazing of tussock grass has been identified in terms of physiography and this follows a similar sequence to the irruptive stage of each herd.

The model of population irruption derived for introduced thar in New Zealand adequately describes the pattern observed amongst the three herds of reindeer on South Georgia. However, the model does not explain the dramatic population crashes seen amongst other introduced reindeer, notably on St Matthew Island. Comparisons between Arctic islands and South Georgia suggest that emigration and climate have probably modified population declines on some islands, but that successful introductions occur only when reindeer depend on vascular species rather than lichens for their winter forage. Studies of introduced island reindeer show the critical influence that aspects of forage quality and availability exert on population dynamics, and suggest that these factors should receive more attention in the controversy over whether continental populations are food- or predator-limited.

The irruptive model also appears to be applicable to the wide range of mammalian introductions that have been made to other far southern islands. Animals ranging in size from mice to cattle and horses, and in feeding habits from herbivores to carnivores, have had a noticeable impact on the floras and faunas of these islands. These introductions provide many as yet untapped possibilities for future research. The relationship between overgrazing and possible exacerbation of erosion remains controversial. Trampling by reindeer is known to affect invertebrate populations, but nothing is known of the effects of introduced herbivore dung on nutrient recycling. The genetics of southern island ungulate populations remain almost unstudied, despite the obvious advantages of isolated herds derived from small numbers of founders.

At present no management plans exist for the introduced reindeer populations on either South Georgia or Iles Kerguelen. Whilst reindeer have had a noticeable impact on native plant communities, it appears that no plant species has been eradicated by uncontrolled grazing. Furthermore, exclusion experiments have shown that, for the most part, native vegetation communities can recover if grazing pressure is removed. However, some long-term and irreversible changes may already have resulted, even if reindeer are now eradicated. This provides a classic management conundrum which must be clarified shortly. With the recognition of the conservation importance of far southern islands, management plans for some are including the active eradication of introduced species. For reindeer, this option must be weighed against their scientific interest.

From this volume it is clear than man holds the key to the future of most reindeer and caribou populations. The herding of northern reindeer and hunting of caribou is an important feature of the Arctic economy. However, reindeer herding now faces an immediate threat resulting from the nuclear accident at the Chernobyl power station in the Soviet Union. Radioactive pollution has become concentrated in the lichen–reindeer food chain, resulting in the condemnation of large numbers of reindeer as unfit for human consumption. The accumulation of long-lived isotopes such as caesium and strontium in the reindeer food chain may even place a longer-term question mark over the future of Eurasian reindeer herding. Hence the small herds on southern islands, which rely on vascular plants less prone to radioactive pollution, may provide even more important control and comparative research material for northern herds in the future.

References

Abbott, I. (1974). Numbers of plant, insect and land bird species on nineteen remote islands in the Southern Hemisphere. *Biological Journal of Linnaean Society*, **6**, 143–152.

Adams, J. (1980). High sediment yields from major rivers of the western Southern Alps, New Zealand. *Nature, London*, **287**, 88–89.

Albon, S.D., Mitchell, B., Huby, B.J. & Brown, D. (1986). Fertility in female red deer (*Cervus elaphus*): the effects of body composition, age and reproductive status. *Journal of Zoology*, London, **209A**, 447–460.

Aleksandrova, V.D. (1937). [On the winter feeding of domesticated reindeer on Novaya Zemlya.] *Sovetskoye olenevodstvo*, **9**, 127–139. (In Russian.)

Alexander, R.D., Hoogland, J.L., Howard, R.D., Noonan, M. & Sherman, P.W. (1979). Sexual dimorphisms, and breeding systems in pinnipeds, ungulates, primates and humans. In *Evolutionary Biology and Human Social Behaviour, an anthropological perspective*, ed. N.A. Chagnon & W. Irons, pp. 402–604. Massachusetts: Duxbury Press.

Allen, H.T. (1920). *Fauna of the Dependencies of the Falkland Islands: Report of the Interdepartmental Committee on Research and Development in the Dependencies of the Falkland Islands*. London: HMSO.

Allen, S.E., Grimshaw, H.M., Parkinson, J.A. & Quarmby, C. (1974). *Chemical Analysis of Ecological Materials*. Oxford: Blackwell Scientific Publications.

Andersen, J. (1953). Analysis of a Danish roe deer population (*Capreolus capreolus*) based upon the extermination of the total stock. *Danish Review of Game Biology*, **2**, 127–155.

Anderson, A.E., Medin, D.E. & Bowden, D.C. (1974). Growth and morphometry of the carcass, selected bones, organs, and glands of mule deer. *Wildlife Monographs*, **39**, 1–122.

Anderson, A.E., Snyder, W.A. & Brown, G.W. (1965). Stomach analyses related to condition in mule deer, Guadalupe Mountains, Mexico. *Journal of Wildlife Management*, **29**, 352–366.

Anderson, R.C. (1965). An examination of wild moose exhibiting neurologic signs in Ontario. *Canadian Journal of Zoology*, **43**, 635–639.

Anderson, R.C. (1971). Neurologic disease in reindeer (*Rangifer tarandus tarandus*) introduced into Ontario. *Canadian Journal of Zoology*, **49**, 159–166.

Anderson, R.C. (1972). The ecological relationships of meningeal worm and native cervids in North America. *Journal of Wildlife Diseases*, **8**, 304–310.

Anderson, R.C. (1976). The impact of parasitic diseases on wildlife populations: Helminths. In *Wildlife Diseases*, ed. L.A. Page, pp. 35–43. New York: Plenum Press.

Anderson, R.M. (1982). *Population Dynamics of Infectious Disease*. London: Chapman & Hall.

Anderson, R.M. & May, R.M. (1978). Regulation and stability of host-parasite population interactions. I. Regulatory processes. *Journal of Animal Ecology*, **47**, 219–247.

Andreev, V.N. (1977). Reindeer pastures in the subarctic territories of theUSSR. In *Handbook of Vegetation Science*, vol. 13, ed. W. Krause, pp. 275–313. The Hague: Junk.

Anon. (1970). The Swiss Rammsonde. *Special Technical Report*, **1**. Illinois: Testlab, Division of GDI, Inc.

Anon. (1971). *Manual of Veterinary Parasitological Laboratory Techniques*. London: HMSO.

Anon. (1983). *Management Plan for the Campbell Islands Nature Reserve*. Wellington: Department of Lands and Survey.

Arnold, G.W. (1964). Factors within plant associations affecting the behaviour and performance of grazing mammals. In *Grazing in Terrestrial and Marine Environments*, ed. D.J. Crisp, pp. 133–154. Oxford: Blackwell Scientific Publications.

Atkinson, I.A.E. (1985). The spread of commensal species of *Rattus* to oceanic islands and their effects on native avifaunas. In *Conservation of Island Birds*, ed. P.J. Moors, pp. 35–81. Cambridge: International Council for Bird Preservation.

Bahn, P.G. (1977). Seasonal migration in southwest France during the late glacial period. *Journal of Archeological Science*, **4**, 245–248.

Baker, J.R. (1938). Evolution of breeding seasons. In *Evolution: Essays on Aspects of Evolutionary Biology Presented to Professor E.S. Goodrich*, ed. G. de Beer, pp. 161–177. Oxford: Clarendon Press.

Ballou, J. & Ralls, K. (1982). Inbreeding and juvenile mortality in small populations of ungulates: a detailed analysis. *Biological Conservation*, **24**, 239–272.

Bandy, P.J., Cowan, I. Mc T. & Wood, A.J. (1970). Comparative growth in four races of black-tailed deer. Part I: Growth in body weight. *Canadian Journal of Zoology*, **48**, 1401–1410.

Banfield, A.W.F. (1954). Preliminary investigation of the barren-ground caribou. *Canadian Wildlife Service, Wildlife Management Bulletin*, Series 1, 10B, 1–112.

Banfield, A.W.F. (1961). A revision of the reindeer and caribou, genus *Rangifer*. *National Museum of Canada, Bulletin*, **177**, 1–137.

Banfield, A.W.F. (1977). *The Mammals of Canada*. Toronto: Toronto University Press.

Bannikov, A.G., Zhirmov, L.S., Lebedeva, & Fandeev, A.A. (1967). *Biology of the Saiga*. Springfield, Virginia: US Department of Commerce.

Baskin, L.M. (1970). [*Northern Reindeer: ecology and behaviour.*] Moscow: Nauka. (In Russian.)

Baskin, L.M. (1983). The causes of calf reindeer mortality. *Acta Zoologica Fennica*, **175**, 133–134.

Baskin, L.M. (1986). Differences in the ecology and behaviour of reindeer populations in the USSR. *Rangifer*, Special Issue **1**, 333–340.

Beaglehole, J.C. (1961). *The Journals of Captain James Cook on His Voyage of Discovery*. 2. *The Voyage of the Resolution and Adventure, 1772–1775*. Cambridge: Cambridge University Press.

Bell, B.G. (1982). Notes on the alien vascular flora of Ile de Possession, Iles Crozet. *Comité National Français des Recherches Antarctiques*, **51**, 325–331.

Bell, C.M., Mair, B.F. & Storey, B.C. (1977). The geology of part of an island arc-marginal basin system in southern South Georgia. *British Antarctic Survey Bulletin*, **46** 109–127.

Bergerud, A.T. (1964a). Relation of mandible length to sex in Newfoundland caribou. *Journal of Wildlife Management*, **28**, 54–56.

Bergerud, A.T. (1964b). Field determination of annual parturition in Newfoundland caribou. *Journal of Wildlife Management*, **28**, 477–480.

Bergerud, A.T. (1971). The population dynamics of Newfoundland caribou. *Wildlife Monographs*, **25**, 1–55.

Bergerud, A.T. (1972). Food habits of Newfoundland caribou. *Journal of Wildlife Management*, **36**, 913–923.

Bergerud, A.T. (1973). Movement and rutting behaviour of caribou (*Rangifer tarandus*) at Mount Albert, Quebec. *Canadian Field-Naturalist*, **87**, 357–369.

Bergerud, A.T. (1974a). Rutting behaviour of Newfoundland caribou. In *TheBehaviour of Ungulates and its Relation to Management*, ed. V. Geist & F. Walther, pp. 395–435. Morges: IUCN.

Bergerud, A.T. (1974b). The role of the environment in the aggregation, movement and disturbance behaviour of caribou. In *The Behaviour of Ungulates and its Relation to Management*, ed. V. Geist & F. Walther, pp. 552–584. Morges: IUCN.

Bergerud, A.T. (1974c). Decline of caribou in North America following settlement. *Journal of Wildlife Management*, **38**, 757–770.

Bergerud, A.T. (1975). The reproductive season of Newfoundland caribou. *Canadian Journal of Zoology*, **53**, 1213–1221.

Bergerud, A.T. (1976). The annual antler cycle in Newfoundland caribou. *Canadian Field-Naturalist*, **90**, 449–463.

Bergerud, A.T. (1980a). A review of the population dynamics of caribou and wild reindeer in North America. In *Proceedings of 2nd International Reindeer/Caribou Symposium*, ed. E. Reimers, E. Gaare & S. Skjenneberg, pp. 556–581. Trondheim: Direktoratet for Vilt og Ferskvannsfisk.

Bergerud, A.T. (1980b). Status of *Rangifer* in Canada. 1. Woodland caribou (*Rangifer tarandus caribou*). In *Proceedings of 2nd International Reindeer/Caribou Symposium*, ed. E. Reimers, E. Gaare & S. Skjenneberg, pp.748–753. Trondheim: Direktoratet for Vilt og Ferskvannsfisk.

Bergerud, A.T. (1983). The natural population control of caribou. In *Symposium on Natural Regulation of Wildlife Populations*, ed. F.L. Bunnell, D.S. Eastmann & J.M. Peek, pp. 14–61. Moscow: University of Idaho Press.

Bergerud, A.T., Jakimchuk, R.D. & Carruthers, D.R. (1984). The buffalo of the north: caribou (*Rangifer tarandus*) and human developments. *Arctic*, **37**, 7–22.

Bergerud, A.T. & Russell, H.L. (1964). Evaluation of rumen food analysis for Newfoundland caribou. *Journal of Wildlife Management*, **28**, 809–814.

Bergerud, A.T., Nolan, M.J., Curnew, R. & Mercer, W.E. (1983). Growth of the Avalon Peninsula, Newfoundland caribou herd. *Journal of Wildlife Management*, **47**, 989–998.

Berry, R.J., Bonner, W.N. & Peters, J. (1979). Natural selection in house mice (*Mus musculus*) from South Georgia (South Atlantic Ocean). *Journal of Zoology, London*, **189**, 385–398.

Berry, R.J., Peters, J. & Van Aarde, R.J. (1978). Subantarctic house mice: colonisation, survival and selection. *Journal of Zoology, London*, **184**, 127–141.

Bjarghov, R.S, Fjellheim, P., Hove, K., Jacobsen, E., Skjenneberg, S. & Try, K. (1976). Nutritional effects on serum enzymes and other blood constituents in reindeer calves (*Rangifer tarandus tarandus*). *Comparative Biochemistry and Physiology*, **55A**, 187–193.

Bjarghov, R.S., Jacobsen, E. & Skjenneberg, S. (1977). Composition of liver, bone and bone-marrow of reindeer (*Rangifer tarandus tarandus*) measured at two different seasons of the year. *Comparative Biochemistry and Physiology*, **56A**, 337–341.

Block, W. (1984). Terrestrial microbiology, invertebrates and ecosystems. In *Antarctic Ecology*, ed. R.M. Laws, pp. 163–236. London: Academic Press.

Blood, D.C. & Henderson, J.A. (1979). *Veterinary Medicine*, 5th edn. London: Balliere, Tindall & Cassell.

Bloomfield, M. (1980). Patterns of seasonal habitat selection exhibited by mountain caribou in central British Columbia, Canada. In *Proceedings of 2nd International Reindeer/Caribou Symposium*, ed. E. Reimers, E. Gaare & S. Skjenneberg, pp. 10–18. Trondheim: Direktoratet for Vilt og Ferskvannsfisk.

Bobek, B. (1977). Summer food as the factor limiting roe deer population size. *Nature, London*, **268**, 47–49.

Bonner, W.N. (1958). The introduced reindeer of South Georgia. *Falkland Island Dependencies Survey, Scientific Report*, **22**, 1–8.

Bonner, W.N. (1982). *Seals and man: a study of interactions*. Seattle: University of Washington Press.

Bonner, W.N. (1984a). Introduced mammals. In *Antarctic Ecology*, ed. R.M. Laws, pp. 237–278. London: Academic Press.

Bonner, W.N. (1984b). Conservation in the Antarctic. In *Antarctic Ecology*, ed. R.M. Laws, pp. 821–850. London: Academic Press.

Bonner, W.N. (1985). Impact of fur seals on the terrestrial environment at South Georgia. In *Antarctic Nutrient Cycles and Food Webs*, ed. W.R. Siegfried, P.R. Condy & R.M. Laws, pp. 641–646. Berlin: Springer-Verlag.

Bonner, W.N. & Leader-Williams, N. (1977). House mice on South Georgia. *Polar Record*, **18**, 512.

Borg, K. (1970). On mortality and reproduction of the roe deer in Sweden during the period 1948-1969. *Viltrevy*, **7**, 121–146.

Borozdin, E.K. (1969). [Histological structure of ovaries and development of oocytes in reindeer]. *Trudy Nauchno-issledovatel'skogo*, **17**, 85–93. (In Russian.)

Borzjonov, B.B., Pavlov, V.V. & Yakushkin, G.D. (1979). Adaptations of the Taimyr wild reindeer population. In *Abstracts of 2nd International Reindeer/Caribou Symposium*, pp. 60. Røros, Norway 1979.

Bousses, P., Chapuis, J.L. & Arthur, C.R. (1988). Données préliminaires sur la biologie et l'écologie du lapin (*Oryctolagus cuniculus*) des Iles Kerguelen. *Comité National Français des Recherches Antarctiques*, in press.

Brody, S. (1945). *Bioenergetics and Growth, With Special Reference to the Efficiency Complex in Domestic Animals*. New York: Reinhold Publishing Corporation.

Brothers, N.P., Copson, G.R., Skira, I.J. & Eberhard, E. (1982). Control of rabbits on Macquarie Island by myxomatosis. *Australian Journal of Wildlife Research*, **9**, 477–485.

Bubenik, A.B. (1975a). Taxonomic value of antlers in genus *Rangifer*, H. Smith. *Biological Papers of University of Alaska, Special Report*, **1**, 41–63.

Bubenik, A.B. (1975b). Significance of antlers in the social life of barren-ground caribou. *Biological Papers of University of Alaska, Special Report*, **1**, 436–461.

Bunnell, F.L. (1982). The lambing period of mountain sheep: synthesis, hypotheses and tests. *Canadian Journal of Zoology*, **60**, 1–14.

Burch, E.S. (1975). The caribou/wild reindeer as a human resource. *American Antiquity*, **37**, 339–368.

Burger, A.E. (1985). Terrestrial food-webs in the sub-Antarctic: island effects. In *Antarctic Nutrient Cycles and Food Webs*, ed. W.R. Siegfried, P.R. Condy & R.M. Laws, pp. 582–591. Berlin: Springer-Verlag.

Bye, K. (1985). Cestodes of reindeer (*Rangifer tarandus platyrhynchus* Vrolik) on the Arctic islands of Svalbard. *Canadian Journal of Zoology*, **63**, 2885–2887.

Bye, K. & Halvorsen, O. (1983). Abomasal nematodes of the Svalbard reindeer(*Rangifer tarandus platyrhynchus* Vrolik). *Journal of Wildlife Diseases*, **19**, 101–105.

Calef, G.W. (1980). Status of *Rangifer* in Canada. II. Status of *Rangifer* in the Northwest Territories. In *Proceedings of 2nd International Reindeer/Caribou Symposium*, ed. E. Reimers, E. Gaare & S. Skjenneberg, pp. 754–759. Trondheim: Direktoratet for Vilt og Ferskvannsfisk.

Callaghan, T.V. Lewis Smith, R.I. & Walton, D.W.H. (1976). The IBP Bipolar Botanical Project. *Philosophical Transactions of Royal Society of London*, Series B, **274**, 315–319.

Campbell, D.J. & Rudge, M.R. (1984). Vegetation changes induced over ten years by goats and pigs at Port Ross, Auckland Islands (Subantarctic). *New Zealand Journal of Ecology*, **7**, 103–118.

Caughley, G. (1966). Mortality patterns in mammals. *Ecology*, **47**, 906–918.

Caughley, G. (1970). Eruption of ungulate populations, with emphasis on Himalayan thar in New Zealand. *Ecology*, **51**, 53–72.

Caughley, G. (1971). Season of births for northern-hemisphere ungulates in New Zealand. *Mammalia*, **35**, 204–219.

Caughley, G. (1976a). Wildlife management and the dynamics of ungulate populations. In *Applied Biology*, vol. I, ed. T.H. Coaker, pp. 183–246. London: Academic Press.

Caughley, G. (1976b). Plant-herbivore systems. In *Theoretical Ecology*, ed. R.M. May, pp. 94–113. Oxford: Blackwell Scientific Publications.

Caughley, G. (1977). *Analysis of Vertebrate Populations*. London: Wiley.

Caughley, G. (1979). What is this thing called carrying capacity? In *North Americn Elk: Ecology, Behaviour and Management*, ed. M.S. Boyce, L.D. Hayden-Wing, pp. 2–8. Laramie: University of Wyoming Press.

Caughley, G. (1981). Overpopulation. In *Problems in Management of Locally Abundant Wild Mammals*, ed. P.A. Jewell & S. Holt, pp. 7–19. New York: Academic Press.

Challies, C.N. (1975). Feral pigs (*Sus scrofa*) on Auckland Island: status, and effects on vegetation and nesting sea birds. *New Zealand Journal of Zoology*, **2**, 479–490.

Chaplin, R.E. & White, R.W.G. (1972). The influence of age and season on the activity of the testes and epididymides of the fallow deer, *Dama dama*. *Journal of Reproduction and Fertility*, **30**, 361–369.

Chapman, D.I. (1975). Antlers – bones of contention. *Mammal Review*, **5**, 121–172.

Chapman, D.I. & Chapman, N. (1975). *Fallow Deer: their history, distribution and biology*. Lavenham, England: Terence Dalton.

Charlesworth, J.K. (1957). *The Quaternary Era with special reference to its glaciation*. London: Edward Arnold.

Cheatum, E.L. & Severinghaus, C.W. (1950). Variations in fertility of white-tailed deer related to range conditions. *Transactions of North American Wildlife and Natural Resources Conference*, **15**, 170–189.

Clapperton, C.M. (1971). Geomorphology of the Stromness Bay–Cumberland Bay area, South Georgia. *British Antarctic Survey, Scientific Reports*, **70**, 1–25.

Clark, M.R. & Dingwall, P.R. (1985). *Conservation of islands in the Southern Ocean: a review of the protected areas of Insulantarctica*. Gland: International Union for Conservation of Nature and Natural Resources.

Clausen, B., Dam, A., Elvestad, K., Krogh, H.V. & Thing, H. (1980). Summer mortality among caribou calves in west Greenland. *Nordisk Veterinaermedicin*, **32**, 291–300.

Clayton, R.A.S. (1982). A preliminary investigation of the geochemistry of greywackes from South Georgia. *British Antarctic Survey Bulletin*, **51**, 89–109.

Clutton-Brock, J. (1977). Man-made dogs. *Science*, **197**, 1340–1342.

Clutton-Brock, T.H. (1974). Primate social organisation and ecology. *Nature, London*, **250**, 539–542.

Clutton-Brock, T.H. (1982). The functions of antlers. *Behaviour*, **79**, 109–125.

Clutton-Brock, T.H., Albon, S.D. & Guinness, F.E. (1985b). Parental investment and sex differences in juvenile mortality in birds and mammals. *Nature, London*, **313**, 131–133.

Clutton-Brock, T.H., Albon, S.D. & Harvey, P.H. (1980). Antlers, body size and breeding group size in the Cervidae. *Nature, London*, **285**, 565–566.

Clutton-Brock, T.H., Guinness, F.E. & Albon, S.D. (1982). *Red deer: behavior and ecology of two sexes*. Chicago: Chicago University Press.

Clutton-Brock, T.H., Major, M. & Guinness, F.E. (1985a). Population regulation in male and female red deer. *Journal of Animal Ecology*, **54**, 831–846.

Corker, C.S. & Davidson, D.W. (1978). The radioimmunoassay of testosterone in various biological fluids without chromatography. *Journal of Steroid Biochemistry*, **9**, 373–374.

Costin, A.B. & Moore, D.M. (1960). The effects of rabbit grazing on the grasslands of Macquarie Island. *Journal of Ecology*, **48**, 729–739.

Courtright, A.M. (1959). Results of some detailed analyses of caribou rumen contents. *Proceedings of Alaskan Scientific Conference*, **10**, 28–35.

Crête, M. & Jordan, P.A. (1981). Régime alimentaire des orignaux du sud-ouest Québécois pour les mois d'Avril à Octobre. *Canadian Field-Naturalist*, **95**, 50–56.

Crook, J.H. & Gartlan, J.S. (1966). Evolution of primate societies. *Nature, London*, **210**, 1200–1203.

Croxall, J.P. (1984). Seabirds. In *Antarctic Ecology*, ed. R.M. Laws, pp. 533–619. London: Academic Press.

Croxall, J.P., McInnes, S.J. & Prince, P.A. (1984a). The status and conservation of seabirds at the Falkland Islands. In *Status and Conservation of the World's Seabirds*, ed. J.P. Croxall, P.G.H. Evans & R.W. Schrieber, pp. 271–291. Cambridge: International Council for Bird Preservation.

Croxall, J.P., Prince, P.A., Hunter, I., McInnes, S.J. & Copestake, P.G. (1984b). The seabirds of the Antarctic Peninsula, islands of the Scotia Sea, and Antarctic Continent between 80°W and 20°W: their status and conservation. In *Status and Conservation of the World's Seabirds*, ed. J.P. Croxall, P.G.H. Evans & R.W. Schrieber, pp. 637–666. Cambridge: International Council for Bird Preservation.

Danilov, P.I. & Markovsky, V.A. (1983). Forest reindeer (*Rangifer tarandus fennicus* Lönnb.) in Karelia. *Acta Zoologica Fennica*, **175**, 33–34.

Darby, W.R. & Duquette, L.S. (1986). Woodland caribou and forestry in Northern Ontario, Canada. *Rangifer*, Special Issue **1**, 87–93.

Darwin, C. (1859). *On the Origin of Species*. London: John Murray.

Dauphiné, T.C. (1975). Kidney weight fluctuations affecting kidney fat index of caribou. *Journal of Wildlife Management*, **39**, 379–386.

Dauphiné, T.C. (1976). Biology of the Kaminuriak population of barren-ground caribou. Part 4. Growth, reproduction and energy reserves. *Canadian Wildlife Service, Report Series*, **38**, 1–69.

Dauphiné, T.C. (1978). Morphology of barren-ground caribou ovaries. *Canadian Journal of Zoology*, **56**, 1684–1696.

Dauphiné, T.C. & McClure, R.L. (1974). Synchronous mating in Canadian barren-ground caribou. *Journal of Wildlife Management*, **38**, 54–66.

Davies, D.V., Mann, T. & Rowson, L.E.A. (1957). Effect of nutrition on the onset of male sex hormone activity and sperm formation in monozygous bull calves. *Proceedings of Royal Society of London*, Series B, **147**, 332–351.

Davies, W. (1939). *The grasslands of the Falkland Islands: Report to the Crown Agents and Falkland Islands Colonial Secretary*. Stanley, Falkland Islands: Government Printer.

Davis, J.L. (1980). Status of *Rangifer* in USA. In *Proceedings of 2nd International Reindeer/Caribou Symposium*, ed. E. Reimers, E. Gaare & S. Skjenneberg, pp. 793–797. Trondheim: Direktoratet for Vilt og Ferskvannsfisk.

Davis, T.A. (1973). Asymmetry of reindeer antlers. *Forma et Functio*, **6**, 373–382.

Deardon, B.L., Hansen, R.M. & Pegau, R.E. (1975). Precision of microhistological estimates of ruminant food habits. *Journal of Wildlife Management*, **39**, 402–407.

de Bie, S. (1976). Survivorship in the Svalbard reindeer (*Rangifer tarandus platyrhynchus*) on Edgeøya, Svalbard. *Norsk Polarinstitutt Årbok 1976*, 249–270.

Derenne, P. (1976). Notes sur la biologie du chat haret de Kerguelen. *Mammalia*, **40**, 531–595.

Derenne, P. & Mougin, J.L. (1976). Données écologiques sur les mammifères introduits de l'Ile aux Cochons, Archipel Crozet (46°06'S, 50°14'E). *Mammalia*, **40**, 21–53.

Dilks, P.J. (1979). Observations on the food of feral cats on Campbell Island. *New Zealand Journal of Ecology*, **2**, 64–66.

Dilks, P.J. & Wilson, P.R. (1979). Feral sheep and cattle and Royal albatrosses on Campbell Island; population trends and habitat changes. *New Zealand Journal of Zoology*, **6**, 127–139.

Dingwall, P.R. (1988). Conservation of New Zealand's southern outlying nature reserves: achievements and future needs. *Biological Conservation*, in press.

Doerr, J.G. & Dieterich, R.A. (1979). Mandibular lesions in the Western Arctic caribou herd of Alaska. *Journal of Wildlife Diseases*, **15**, 309–318.

Dorst, J. & Milon, Ph. (1964). Acclimation et conservation de la nature dans les iles subantarctiques françaises. In *Biologie Antarctique*, ed. R. Carrick, M.W. Holdgate & J. Prévost, pp. 579–588. Paris: Hermann.

Dott, H.M. & Utsi, M.N.P. (1973). Artificial insemination of reindeer (*Rangifer tarandus*). *Journal of Zoology, London*, **170**, 505–508.

Dreux, P. (1974). The cat population of Péninsule Courbet, Iles Kerguelen: an example of the founder effect. *Polar Record*, **17**, 53–54.

Dunbar, M.J. (1977). The evolution of polar ecosystems. In *Adaptations within Antarctic Ecosystems*, ed. G.A. Llano, pp. 1063–1076. Washington, DC: Smithsonian Institution.

Dunn, A.M. (1968). The wild ruminant as reservoir host of helminth infection. *Symposium of Zoological Society of London*, **24**, 221–248.

Dunn, A.M. (1978). *Veterinary Helminthology*, 2nd edn. London: Heinemann.

Dzieciolowski, R. (1970). Food of the red deer as determined by rumen content analysis. *Acta Theriologica*, **15**, 89–110.

Edwards, R.Y. (1956). Snow depths and ungulate abundance in the mountains of western Canada. *Journal of Wildlife Management*, **20**, 159–168.

Edwards, R.Y. & Ritcey, R.W. (1960). Foods of caribou in Wells Grey Park, British Columbia. *Canadian Field–Naturalist*, **74**, 3–7.

Egilsson, K. (1983). [*The food and pasture of reindeer in Iceland.*] Reykjavik: Orkustofnun. (In Icelandic.)

Eloranta, E. & Nieminen, M. (1986). Calving of the experimental reindeer herd in Kaamanen during 1970–85. *Rangifer*, Special Issue **1**, 115–121.

Elton, C.S. (1942). *Voles, Mice and Lemmings: problems in population dynamics*. London: Oxford University Press.

Elton, C.S. (1958). *The Ecology of Invasions by Animals and Plants*. London: Methuen & Co. Ltd.

Espmark, Y. (1964a). Rutting behaviour in reindeer (*Rangifer tarandus* L.). *Animal Behaviour*, **12**, 159–163.

Espmark, Y. (1964b). Studies in dominance-subordination relationships in a group of semi-domestic reindeer (*Rangifer tarandus* L.). *Animal Behaviour*, **12**, 420–426.

Espmark, Y. (1971a). Mother–young relationship and ontogeny of behaviour in reindeer (*Rangifer tarandus* L.). *Zeitschrift für Tierpsychologie*, **29**, 42–81.

Espmark, Y. (1971b). Antler shedding in relation to parturition in female reindeer. *Journal of Wildlife Management*, **35**, 175–177.

Estes, R.D. (1966). Behaviour and life history of the wildebeest, (*Connochaetes taurinus* Burchell). *Nature, London*, **212**, 999–1000.

Falkland Islands Dependency of South Georgia File No. 650. Deposited at Scott Polar Research Institute, Cambridge. Archive Number MS 1228/-.

Finney, D.J. (1964). *Statistical Method in Biological Assay*. London: Charles Griffin.

Flerov, K.K. (1960). *Fauna of USSR* Vol. 1. *Mammals: Musk deer and deer*. Jerusalem: Israel Program for Scientific Translations.

Fletcher, T.J. (1974) The timing of reproduction in red deer (*Cervus elaphus*) in relation to latitude. *Journal of Zoology, London*, **172**, 363–367.

Fletcher, T.J. (1978). The induction of male sexual behaviour in red deer (*Cervus elaphus* L.) by the administration of testosterone to hinds and estradiol-17β– to stags. *Hormones and Behaviour*, **11**, 74–88.

Flook, D.R. (1970). Causes and implications of an observed sex differential in the survival of wapiti. *Canadian Wildlife Service, Report Series*, **11**, 1–71.

Foley, R.A. & Atkinson, S. (1984). A dental abnormality amongst a population of Defassa waterbuck (*Kobus defassa* Ruppell 1835). *African Journal of Ecology*, **22**, 289–294.

Formosov, A.N. (1964). *Snow cover as an Integral Factor of the Environment and its Importance in the Ecology of Mammals and Birds*. Edmonton: University of Alberta, Occasional Publication No. 1, Boreal Institute.

Fowler, C.W. (1987). A review of density dependence in populations of large mammals. *Current Mammalogy*, **1**, 401–441.

Franzmann, A.W., LeResche, R.E., Rausch, R.A. & Oldmeyer, J.L. (1978). Alaskan moose measurements and weights and measurement–weight relationships. *Canadian Journal of Zoology*, **56**, 298–306.

French, D.D. & Smith, V.R. (1985). A comparison between Northern and Southern Hemisphere tundras and related ecosystems. *Polar Biology*, **5**, 5–21.

Fuller, T.K. & Keith, L.B. (1981). Woodland caribou population dynamics in northeastern Alberta. *Journal of Wildlife Management*, **45**, 197–213.

Gaare, E., Sørenson, A. & White, R.G. (1977). Are rumen samples representative of the diet? *Oikos*, **29**, 390–395.

Gasaway, W.C. & Coady, J.W. (1974). Review of energy requirements and rumen fermentation in moose and other ruminants. *Naturaliste Canadien*, **101**, 227–262.

Gasaway, W.C., Stephenson, R.O., Davis, J.L., Shepherd, P.E.K. & Burris, O.E. (1983). Interrelationships between wolves, prey and man in interior Alaska. *Wildlife Monographs*, **84**, 1–50.

Gates, C.C., Adamczewski, J.Z. & Mulders, R. (1986). Population dynamics, winter ecology, and social organisation of Coats Island caribou. *Arctic*, **39**, 216–222.

Gauthier, D.A. & Theberge, J.B. (1986). Wolf predation in the Burwash caribou herd, southwest Yukon. *Rangifer*, Special Issue **1**, 137–144.

Gebczynska, Z. (1980). Food of the roe deer and red deer in the Bialowieza Primeval Forest. *Acta Theriologica*, **25**, 487–500.

Geist, V. (1971). *Mountain Sheep: A study in behaviour and evolution*. Chicago: Chicago University Press.

Geller, M. Kh. & Borzhanov, B.B. (1984). Migrations and seasonal distribution of reindeer populations of Taimyr. In *Wild Reindeer of the Soviet Union*, ed. E.E. Syroechkovskii, pp. 71–79. Washington, DC: United States Department of the Interior.

Gleeson, J.P. & van Rensburg, P.J.J. (1982). Feeding ecology of the house mouse, *Mus musculus*, on Marion Island. *South African Journal of Antarctic Research*, **12**, 34–39.

Godkin, G.F. (1986). Fertility and twinning in Canadian reindeer. *Rangifer*, Special Issue **1**, 145–150.

Golley, F.B. (1957). An appraisal of ovarian analysis in determining reproductive performance of black-tailed deer. *Journal of Wildlife Management*, **21**, 62–65.

Gorbunov, E.I. (1939). [Sexual cycle and methods of diagnosis of rut of reindeer does.] *Trudy Nauchno-issledovatel'skogo*, **5**, 101–134. (In Russian.)

Goss, R.J. (1980). Is antler asymmetry in reindeer and caribou genetically determined? In *Proceedings of 2nd International Reindeer/Caribou Symposium*, ed. E. Reimers, E. Gaare & S. Skjenneberg, pp. 364–372. Trondheim: Direktoratet for Vilt og Ferskvannsfisk.

Goss, R.J. (1983). *Deer Antlers: Regeneration, function and evolution*. New York: Academic Press.

Goss, R.J., Severinghaus, C.W. & Free, S. (1964). Tissue relationships in the development of pedicles and antlers in the Virginia deer. *Journal of Mammalogy*, **45**, 61–68.

Gossow, H. (1974). Natural mortality pattern in the Spitzbergen reindeer. *Proceedings of International Congress of Game Biology*, **11**, 103–106.

Greene, S.W. (1964). The vascular flora of South Georgia. *British Antarctic Survey Scientific Reports*, **45**, 1–58.

Grenfell, W.T. (1934). *Forty Years for Labrador*. London: Hodder & Stoughton.

Gressitt, J.L. (ed.) (1970). *Subantarctic entomology, particularly of South Georgia and Heard Island. Pacific Insects Monograph*, Vol. 23. Hawaii: Bishop Museum.

Griffen, B.J. (1980). Erosion and rabbits on Macquarie Island: some comments. *Papers and Proceedings of the Royal Society of Tasmania*, **114**, 81–83.

Griffiths, G.A. (1979). High sediment yields from major rivers in the western Southern Alps, New Zealand. *Nature, London*, **282**, 61–63.

Groves, E.W. (1981). Vascular plant collections from the Tristan da Cunha group of islands. *Bulletin of the British Museum (Natural History), Botany Series*, **8**, 333–420.

Guinness, F.E., Gibson, R.M. & Clutton-Brock, T.H. (1978a). Calving times of red deer (*Cervus elaphus*) on Rhum. *Journal of Zoology, London*, **185**, 105–114.

Guinness, F.E., Clutton-Brock, T.H. & Albon, S.D. (1978b). Factors affecting calf mortality in red deer (*Cervus elaphus*). *Journal of Animal Ecology*, **47**, 817–832.

Gunn, A., Miller, F.L. & Thomas, D.C. (1981). The current status and future of Peary caribou *Rangifer tarandus pearyi* on the Arctic islands of Canada. *Biological Conservation*, **19**, 283–296.

Gunn, R.G. (1970). A note on the effect of broken mouth on the performance of Scottish blackface hill ewes. *Animal Production*, **12**, 517–520.

Gunn, T.C. & Walton, D.W.H. (1985). Storage carbohydrate production and overwintering strategy in a winter-green tussock grass on South Georgia (Sub Antarctic). *Polar Biology*, **4**, 237–242.

Haber, G.C. & Walters, C.J. (1980). Dynamics of the Alaska-Yukon caribou herds and management implications. In *Proceedings of 2nd International Reindeer/Caribou Symposium*, ed. E. Reimers, E. Gaare & S. Skjenneberg, pp. 645–663. Trondheim: Direktoratet for Vilt og Ferskvannfisk.

Hadwen, S. & Palmer, L.J. (1922). Reindeer in Alaska. *United States Department of Agriculture, Bulletin*, **1089**, 1–74.

Hagen, A. (1965). *Rock carvings in Norway*. Oslo: J.G. Tanum Forlag.

Hakala, A.V.K., Staaland, H., Pulliainen, E. & Røed, K.N. (1985). Taxonomy and history of arctic island reindeer with special reference to Svalbard reindeer. *Aquilo: Seria Zoologica*, **23**, 1–11.

Halls, L.K. (ed.) (1984). *White-tailed Deer: ecology and management*. Harrisburg: Stackpole.

Halvorsen, O., Andersen, J., Skorping, A. & Lorentzen, G. (1980). Infection in reindeer with the nematode *Elaphostrongylus rangiferi* Mitskevich in relation to climate and distribution of intermediate hosts. In *Proceedings of 2nd International Reindeer/Caribou Symposium*, ed. E. Reimers, E. Gaare & S. Skjenneberg, pp. 449–455. Trondheim: Direktoratet for Vilt og Ferskvannsfisk.

Hamilton, W.J. & Blaxter, K.L. (1980). Reproduction in farmed red deer. 1. Hind and stag fertility. *Journal of Agricultural Science*, **95**, 261–273.

Hanks, J. (1981). Characterisation of population condition. In *Population Dynamics of Large Mammals*, ed. C.W. Fowler & T.D. Smith, pp. 47–73. New York: Wiley & Sons.

Hanson, H.C. (1952). Importance and development of the reindeer industry in Alaska. *Journal of Range Management*, 5, 243–251.

Harrison, R.J. (1962). The structure of the ovary in mammals. In *The Ovary*, 1st edn, vol. 1, ed. S. Zuckerman, pp. 143–187. New York: Academic Press.

Headland, R.K. (1984). *The Island of South Georgia*. Cambridge: Cambridge University Press.

Heard, D.C. & Calef, G.W. (1986). Population dynamics of the Kaminuriak caribou herd, 1968–1985. *Rangifer*, Special Issue **1**, 159–166.

Helle, T. (1980). Abundance of warble fly (*Oedemagena tarandi*) larvae in semi-domestic reindeer (*Rangifer tarandus*) in Finland. *Report of Kevo Subarctic Research Station*, **16**, 1–6.

Helle, T. (1981). Studies on wild forest reindeer (*Rangifer tarandus fennicus* Lonn.) and semi-domestic reindeer (*Rangifer tarandus tarandus* L.) in Finland. *Acta Universitatis Oulensis*, Series A, **107**, 1–34.

Helle, T. & Aspi, J. (1983). Effects of winter grazing by reindeer on vegetation. *Oikos*, **40**, 337–343.

Hellesnes, P. (1935). Rensnyltere. *Norsk Veterinaertidsskrift*, **47**, 117–137.

Hemming, J.E. (1975). Population growth and movement patterns of the Nelchina caribou herd. *Biological Papers of University of Alaska, Special Report*, **1**, 162–169.

Henshaw, J. (1968a). A theory for the occurrence of antlers in females of genus *Rangifer*. *Deer*, **1**, 222–26.

Henshaw, J. (1968b). The activities of wintering caribou in Northwestern Alaska in relation to weather and snow conditions. *International Journal of Biometeorology*, **12**, 21–27.

Henshaw, J. (1970). Consequences of travel in the rutting of reindeer and caribou (*Rangifer tarandus*). *Animal Behaviour*, **18**, 256–258.

Herre, W. (1955). *Das Ren als Haustier: Eine zoologische Monographie*. Leipzig: Akademische Verlagsgesellschaft.

Hilborn, R. & Sinclair, A.R.E. (1979). A simulation of the wildebeest population, other ungulates, and their predators. *Serengeti: Dynamics of an Ecosystem*, ed. A.R.E. Sinclair & M. Norton-Griffiths, pp. 287–309. Chicago: Chicago University Press.

Hjeljord, O. (1975). Studies on the Svalbard reindeer. *Norsk Polar Institut Årbok 1973*, 113–123.

Hoeck, H.N. (1984). Introduced faunas. In *Key Environments – Galapagos*, ed R. Perry, pp. 233–245. Oxford: Pergamon Press.

Hofmann, R.R. & Stewart, D.R.M. (1972). Grazer or browser: a classification based on the stomach structure and feeding habits of East African ruminants. *Mammalia*, **36**, 226–240.

Holdgate, M.W. (1965). Fauna of the Tristan da Cuhna Islands. *Philosophical Transactions of Royal Society of London*, Series B, **249**, 361–402.

Holdgate, M.W. (1967). The influence of introduced species on the ecosystems of temperate oceanic islands. *International Union for the Conservation of Nature Publications, New Series*, **9**, 151–176.

Holdgate, M.W. & Wace N.M. (1961). The influence of man on the floras and faunas of Southern islands. *Polar Record*, **10**, 475–494.

Hollander, M. & Wolfe, D.A. (1973). *Non-parametric Statistical Methods*. New York: Wiley & Sons.

Holthe, V. (1975). Calving seasons in different populations of wild reindeer in South Norway. *Biological Papers of University of Alaska, Special Report*, **1**, 194–198.

Houston, D.B. (1982). *The Northern Yellowstone Elk: ecology and management*. New York: Macmillan.

Hove, K. & Jacobsen, E. (1975). Renal excretion of urea in reindeer: effect of nutrition. *Acta Veterinaria Scandanavica*, **16**, 513–519.

Howard, W.E. (1964). Introduced browsing mammals and habitat stability in New Zealand. *Journal of Wildlife Management*, **28**, 421–429.

Hustich, I. (1951). The lichen woodlands in Labrador and their importance as winter pastures for domesticated reindeer. *Acta Geographica, Helsinki*, **12**, 1–48.

Huxley, J.S. (1931). The relative size of antlers in deer. *Proceedings of Zoological Society of London*, **19**, 819–864.

Hyvärinen, H., Helle, T., Väyrynen, R. & Väyrynen, P. (1975). Seasonal and nutritional effects of serum protein and urea concentrations in the reindeer (*Rangifer tarandus tarandus* L.) *Journal of Nutrition*, **33**, 63–72.

Ingold, T. (1980). *Hunters, Pastoralists and Ranchers: reindeer economics and their transformations*. Cambridge: Cambridge University Press.

Ivlev, V.S. (1961). *Experimental Ecology of the Feeding of Fishes*. New Haven: Yale University Press.

Jackson, J.E. (1974). Feeding habits of deer. *Mammal Review*, **4**, 93–101.

Jacobi, A. (1931). *Das Rentier: eine Zoologische Monographie der Gattung Rangifer*. Leipzig: Akademische Verlagsgesellschaft.

Jarman, P.J. (1974). The social organisation of antelope in relation to their ecology. *Behaviour*, **28**, 215–267.

Jenkin, J.F. Johnstone, G.W. & Copson, G.R. (1982). Introduced animal and plant species on Macquarie Island. *Comité National Français des Recherches Antarctiques*, **51**, 301–313.

Jensen, P.V. (1968). Food selection of the Danish red deer (*Cervus elaphus* L.) as determined by examination of rumen contents. *Danish Review of Game Biology*, **5**(3), 1–44.

Johnstone, G.W. (1985). Threats to birds on subantarctic islands. In *Conservation of Island Birds*, ed. P.J. Moors, pp. 101–121. Cambridge: International Council for Bird Preservation.

Jones, E. (1977). Ecology of the feral cat, *Felis catus* (L.) (Carnivora: Felidae) on Macquarie Island. *Australian Wildlife Research*, **4**, 249–262.

Jones, E. & Skira, I.J. (1979). Breeding distribution of the great skua at Macquarie Island in relation to numbers of rabbits. *Emu*, **79**, 19–23.

Jones, R.D. (1966). Raising caribou for an Aleutian introduction. *Journal of Wildlife Management*, **30**, 453–460.

Jordan, P.A., Botkin, D.B. & Wolfe, M.L. (1971). Biomass dynamics in a moose population. *Ecology*, **52**, 147–152.

Jouventin, P., Stahl, J.C., Weimerskirch H. & Mougin, J.L. (1984). The seabirds of the French subantarctic islands and Adélie Land, their status and conservation. In *Status and Conservation of the World's Seabirds*, ed. J.P. Croxall, P.G.H. Evans & R.W. Schreiber, pp. 609–625. Cambridge: International Council for Bird Preservation.

Kay, R.N.B. & Staines, B.W. (1981). The nutrition of red deer (*Cervus elaphus*). *Nutrition Abstracts and Reviews*, **51**, 601–622.

Keage, P.L. (1982). The conservation status of Heard Island and the McDonald Islands. *University of Tasmania Environmental Studies, Occasional Paper*, **13**, 1–100.

Kelsall, J.P. (1968). *The Migratory Barren-ground Caribou of Canada*. Ottawa: Canadian Wildlife Service.

Kelsall, J.P. (1975). Warble fly distribution among some Canadian caribou. *Biological Papers of University of Alaska, Special Report*, **1**, 509–517.

Kelsall, J.P. & Prescott, W. (1971). Moose and deer behaviour in snow in Fundy National Park, New Brunswick. *Canadian Wildlife Service Report Series*, **15**, 1–27.

Kightley, S.P.J. & Lewis Smith, R.I. (1976). The influence of reindeer on the vegetation of South Georgia: I. Long-term effects of unrestricted grazing and the establishment of exclosure experiments in various plant communities. *British Antarctic Survey Bulletin*, **44**, 57–76.

Kischinskii, A.A. (1971). [The modern state of the wild reindeer population on the Novosibirsky Islands.] *Zoologischeskii Zhurnal*, **50**, 117–125. (In Russian.)

Kischinskii, A.A. (1984). Insular populations of wild reindeer in the eastern sector of the Soviet Arctic and methods for their rational exploitation. In *Wild Reindeer of the Soviet Union*, ed. E.E. Syroechkovskii, pp. 158–162. Washington, DC: United States Department of the Interior.

Klein, D.R. (1959). St Matthew Island reindeer-range study. *US Fish and Wildlife Service, Special Scientific Report, Wildlife*, **43**, 1–48.

Klein, D.R. (1962). Rumen contents analysis as an index to range quality. *Transactions of North American Wildlife Conference*, **27**, 150–164.

Klein, D.R. (1968). The introduction, increase, and crash of reindeer on St Matthew Island. *Journal of Wildlife Management*, **32**, 350–367.

Klein, D.R. (1970a). Tundra ranges north of the boreal forest. *Journal of Range Management*, **23**, 8–14.

Klein, D.R. (1970b). Food selection by North American deer and their response to over-utilization of preferred plant species. In *Animal Populations in Relation to their Food Resources*, ed. A. Watson, pp. 25–44. Oxford: Blackwell Scientific Publications.

Klein, D.R. (1980a). Range ecology and management. In *Proceedings of 2nd International Reindeer/Caribou Symposium*, ed. E. Reimers, E. Gaare & S. Skjenneberg, pp. 4–9. Trondheim: Direktoratet for Vilt og Ferskvannsfisk.

Klein, D.R. (1980b). Conflicts between domestic reindeer and their wild counterparts: a review of Eurasian and North American experience. *Arctic*, **33**, 739–756.

Klein, D.R. (1981). The problems of overpopulation of deer in North America. In *Problems in Management of Locally Abundant Wild Mammals*, ed. P.A. Jewell & S. Holt, pp. 119–127. New York: Academic Press.

Klein, D.R. (1982). Fire, lichens and caribou. *Journal of Range Management*, **35**, 390–395.

Klein, D.R. (1986). Latitudinal variation in forage strategies. In *Grazing Research at Northern Latitudes*, ed. O. Gudmundsson, pp. 237–246. New York: Plenum.

Klein, D.R. & Kuzyakin, V. (1982). Distribution and status of wild reindeer in the Soviet Union. *Journal of Wildlife Management*, **46**, 728–733.

Klein, D.R. & Schønheyder, F. (1970). Variation in ruminal nitrogen levels among some Cervidae. *Canadian Journal of Zoology*, **48**, 1437–1442.

Klein, D.R. & Strandgaard, H. (1972). Factors affecting growth and body size of roe deer. *Journal of Wildlife Management*, **36**, 64–79.

Krafft, A. (1981). Villrein i Norge. *Viltrapport*, **18**, 1–92. Trondheim: Direktoratet for Vilt og Ferskvannsfisk.

Krebs, C.J. (1961). Population dynamics of the Mackenzie Delta reindeer herd, 1938–1958. *Arctic*, **14**, 91–100.

Krebs, C.J. & Cowan, I. McT. (1962). Growth studies of reindeer fawns. *Canadian Journal of Zoology*, **40**, 863–869.

Krog, J., Wika, M., Lund-Larsen, T., Nordfjell, J. & Myrnes, I. (1976). Spitzbergen reindeer, *Rangifer tarandus platyrhyncus* Vrolik: Morphology, fat storage, and organ weights in the late winter season. *Norwegian Journal of Zoology*, **24**, 407–417.

Kummeneje, K. (1977). *Dictyocaulus viviparus* infestation in reindeer in northern Norway. *Acta Veterinaria Scandanavica*, **18**, 86–90.

Kuropat, P & Bryant, J.P. (1983). Digestibility of caribou summer forage in arctic Alaska in relation to nutrient, fibre, and phenolic constituents. *Acta Zoologica Fennica*, **175**, 51–52.

Lack, D. (1954). *The Natural Regulation of Animal Numbers*. London: Oxford University Press.

Lack, D. (1976). *Island Biology Illustrated by the Land Birds of Jamaica*. Oxford: Blackwell Scientific Publications.

Lankester, M.W., Crichton, V.J. & Timmermann, H.R. (1976). A protostrongylid nematode (Strongylida: Protostrongylidae) in woodland caribou (*Rangifer tarandus caribou*). *Canadian Journal of Zoology*, **54**, 680–684.

La Perriere, A.J. & Lent, P.C. (1977). Caribou feeding sites in relation to snow characteristics in Northeastern Alaska. *Arctic*, **30**, 101–108.

Laws, R.M. (1953). The elephant seal industry at South Georgia. *Polar Record*, **6**, 746–755.

Laws, R.M. (1981). Large mammal feeding strategies and related overabundance problems. In *Problems in Management of Locally Abundant Wild Mammals*, ed. P.A. Jewell & S. Holt, pp. 217–232. New York: Academic Press.

Laws, R.M. (1984). Seals. In *Antarctic Ecology*, ed. R.M. Laws, pp. 621–715. London: Academic Press.

Laws, R.M. (1985). Animal conservation in the Antarctic. *Symposium of Zoological Society of London*, **54**, 3–23.

Laws, R.M. & Parker, I.S.C. (1968). Recent studies on elephant populations in East Africa. *Symposium of Zoological Society of London*, **21**, 319–359.

Laws, R.M., Parker I.S.C. & Johnstone, R.B. (1975). *Elephants and Their Habitats: the ecology of elephants in North Bunyoro, Uganda.* Oxford: Clarendon Press.

Leader-Williams, N. (1978). The history of the introduced reindeer on South Georgia. *Deer*, **4**, 256–261.

Leader-Williams, N. (1979a). Age determination of reindeer introduced into South Georgia. *Journal of Zoology, London*, **188**, 501–515.

Leader-Williams, N. (1979b). Age-related changes of the testicular and antler cycles of reindeer, *Rangifer tarandus. Journal of Reproduction and Fertility*, **57**, 117–126.

Leader-Williams, N. (1979c). Abnormal testes in reindeer, *Rangifer tarandus. Journal of Reproduction and Fertility*, **57**, 27–130.

Leader-Williams, N. (1980a). Dental abnormalities and mandibular swellings in South Georgia reindeer. *Journal of Comparative Pathology*, **90**, 315–330.

Leader-Williams, N. (1980b). Population dynamics and mortality of reindeer introduced into South Georgia. *Journal of Wildlife Management*, **44**, 640–657.

Leader-Williams, N. (1980c). Observations on the internal parasites of introduced reindeer on South Georgia. *Veterinary Record*, **107**, 393–395.

Leader-Williams, N. (1980d). Population ecology of reindeer on South Georgia. In *Proceedings of 2nd International Reindeer/Caribou Symposium*, ed. E. Reimers, E. Gaare & S. Skjenneberg, pp. 664–676. Trondheim: Direktoratet for Vilt og Ferskvannfisk.

Leader-Williams, N. (1982). Relationship between a disease, host density and mortality in a free-living deer population. *Journal of Animal Ecology*, **51**, 235–240.

Leader-Williams, N. (1985). The sub-Antarctic islands – introduced mammals. In *Key Environments – Antarctica*, ed. W.N. Bonner & D.W.H. Walton, pp. 318–328. Oxford: Pergamon Press.

Leader-Williams, N., Lewis Smith, R.I. & Rothery, P. (1987). Influence of introduced reindeer upon the vegetation of South Georgia: results from a long-term exclusion experiment. *Journal of Applied Ecology*, **24**, 801–822.

Leader-Williams, N. & Payne M.R. (1980). Status of *Rangifer* on South Georgia. In *Proceedings of 2nd International Reindeer/Caribou Symposium*, ed. E. Reimers, E. Gaare & S. Skjenneberg, pp. 786–789. Trondheim: Direktoratet for Vilt og Ferskvannsfisk.

Leader-Williams, N. & Ricketts, C. (1982a). Seasonal and sexual patterns of growth and condition of reindeer introduced into South Georgia. *Oikos*, **38**, 27–39.

Leader-Williams, N. & Ricketts, C. (1982b). Growth and condition of three introduced reindeer herds on South Georgia: the effects of diet and density. *Holarctic Ecology*, **5**, 381–388.

Leader-Williams, N. & Rosser, A.M. (1983). Ovarian characteristics and reproductive performance of reindeer, *Rangifer tarandus. Journal of Reproduction and Fertility*, **67**, 247–256.

Leader-Williams, N., Scott, T.A. & Pratt, R.M. (1981). Forage selection by introduced reindeer on South Georgia, and its consequences for the flora. *Journal of Applied Ecology*, **18**, 83–106.

Lent, P.C. (1965a). Rutting behaviour in a barren-ground caribou population. *Animal Behaviour*, **13**, 259–264.

Lent, P.C. (1965b). Observations of antler shedding by female barren-ground caribou. *Canadian Journal of Zoology*, **43**, 553–558.

Lent, P.C. (1966). Calving and related social behaviour in barren-ground caribou. *Zeitschrift für Tierpsychologie*, **23**, 701–756.

Lent, P.C. (1974). Mother–infant relationships in ungulates. In *The Behaviour of Ungulates and its Relation to Management*, ed. V. Geist & F. Walther, pp. 14–55. Morges: IUCN.

Lent, P.C. & Knutson, D. (1971). Muskox and snow cover on Nunivak Island, Alaska. In *Proceedings of Symposium on Snow and Ice in Relation to Wildlife and Recreation*, ed. A.O. Hauger, pp. 50–62. Ames, Iowa: Iowa State University Press.

Leopold, A. (1943). Wisconsin's deer problem. *Wisconsin Conservation Bulletin*, **8(8)**, 1–11.

Leopold, A.S. & Darling, F.F. (1953). *Wildlife in Alaska: an ecological reconnaissance*. New York: Ronald Press.

LeResche, R.E. (1975). The international herds: present knowledge of the Fortymile and Porcupine caribou herds. *Biological Papers of University of Alaska, Special Report*, **1**, 127–139.

Lesel, R. (1967). Contribution a l'étude écologique de quelques mammifères importés aux Iles Kerguelen. *Terres Australes et Antarctiques Françaises*, **38**, 3–40.

Lesel, R. (1968). Essai d'estimation de la population d'*Oryctolagus cuniculus* L. sur la Peninsule Courbet (Ile de Kerguelen). *Mammalia*, **32**, 612–620.

Lesel, R. (1969). Etude d'un tropeau de bovins sauvages vivant sur l'Ile d'Amsterdam. *Revue d'élevage et de médecine veterinaire de pays tropicaux*, **22**, 107–125.

Lesel, R. & Derenne, P. (1975). Introducing animals to Iles Kerguelen. *Polar Record*, **17**, 485–493.

Lever, C. (1985). *Naturalised Mammals of the World*. Harlow: Longmans.

Lewis Smith, R.I. (1971). An outline of the Antarctic programme of the Bipolar Botanical Project. In *Proceedings IBP Tundra Biome Working Meeting of Analyses of Ecosystems*, Kevo, Finland, September 1970, ed. O.W. Heal, pp. 51–70. Stockholm: International Biological Programme Tundra Biome Steering Committee.

Lewis Smith, R.I. (1984). Terrestrial plant ecology of the sub-Antarctic and Antarctic. In *Antarctic Ecology*, ed. R.M. Laws, pp. 61–162. London: Academic Press.

Lewis Smith, R.I. & Walton, D.W.H. (1973). Calorific value of South Georgia plants. *British Antarctic Survey Bulletin*, **36**, 123–127.

Lewis Smith, R.I. & Walton, D.W.H. (1975). South Georgia: subantarctic. In *Structure and Function of Tundra Ecosystems*, ed. T. Rosswell & O.W. Heal, pp. 399–423. Stockholm: Ecological Bulletin, No. 20.

Lincoln, G.A. (1971a). Puberty in a seasonally breeding male, the red stag deer (*Cervus elaphus* L.). *Journal of Reproduction and Fertility*, **25**, 41–54.

Lincoln, G.A. (1971b). The seasonal reproductive changes in the red deer stag (*Cervus elaphus*). *Journal of Zoology, London*, **163**, 105–123.

Lincoln, G.A. (1979). Appearance of antler pedicles in early foetal life in red deer. *Journal of Embryology and Experimental Morphology*, **29**, 431–437.

Lincoln, G.A. & Kay, R.N.B. (1979). Effects of season on the secretion of LH and testosterone in intact and castrated red deer stags (*Cervus elaphus*). *Journal of Reproduction and Fertility*, **55**, 75–80.

Lincoln, G.A. & Short, R.V. (1980). Seasonal breeding: Nature's contraceptive. *Recent Progress in Hormone Research*, **36**, 1–52.

Lincoln, G.A. Youngson, R.W. & Short, R.V. (1970). The social and sexual behaviour of the red deer stag. *Journal of Reproduction and Fertility*, Supplement 11, 71–103.

Lindgren, E.J. (1935). The reindeer Tungus of Manchuria. *Journal of Royal Central Asian Society*, **22**, 221–231.

Lindgren, E.J. & Utsi, M.N.P. (1948). *Reindeer-breeding in the British Empire*. Unpublished. Mimeo, 28 pp. Deposited at Scott Polar Research Institute, Cambridge.

Lindgren, E.J. & Utsi, V. (1980). Status of *Rangifer* in Britain. In *Proceedings of 2nd International Reindeer/Caribou Symposium*, ed. E. Reimers, E. Gaare & S. Skjenneberg, pp. 744–747. Trondheim: Direktoratet for Vilt og Ferskvannsfisk.

Lindsay, D.C. (1973). Effects of reindeer on plant communities in the Royal Bay area of South Georgia. *British Antarctic Survey Bulletin*, **35**, 101–109.

Lindsay, D.C. (1975). Growth rates of *Cladonia rangiferina* on South Georgia. *British Antarctic Survey Bulletin*, **40**, 40–53.

Linnaeus, C. (1732). *Lachesis Lapponica*. (Translation into English by J.E. Smith, 1811.) London: Richard Taylor & Co.

Linnaeus, C. (1735). *Systema Naturae*. (Translation into English by W. Turton, 1806.) London: Lackington, Allan & Co.

Loudon, A.S.I. (1985). Lactation and neonatal survival of mammals. *Symposium of Zoological Society of London*, **54**, 183–207.

Lovaas, A.L. (1958). Mule deer food habits and range use, Little Belt Mountains, Montana. *Journal of Wildlife Management*, **22**, 257–283.

Lowe, V.P.W. (1969). Population dynamics of the red deer (*Cervus elaphus*) on Rhum. *Journal of Animal Ecology*, **38**, 425–458.

Luick, J.R, Person, S.J., Cameron, R.D. & White, R.G. (1973). Seasonal variations in glucose metabolism of reindeer (*Rangifer tarandus*) estimated with [U^{-14} C] glucose and [3^{-3} H] glucose. *British Journal of Nutrition*, **29**, 245–259.

MacArthur, R.H. & Wilson, E.O. (1967). *The Theory of Island Biogeography*. Princeton: Princeton University Press.

McCullough, D.R. (1969). *The Tule Elk: its history, behaviour and ecology*. Berkeley: University of California Press.

McCullough, D.R. (1979). *The George River Deer Herd: Population ecology of a k-selected species.*. Ann Arbor: University of Michigan Press.

McEwan, E.H. (1963). *Reproduction of barren-ground caribou (Rangifer tarandus groenlandicus Linnaeus) with relation to migration*. Ph.D. thesis, McGill University, Montreal.

McEwan, E.H. (1968). Growth and development of the barren-ground caribou. II. Post-natal growth rates. *Canadian Journal of Zoology*, **46**, 1023–1029.

McEwan, E.H. (1971). Twinning in caribou. *Journal of Mammalogy*, **52**, 479.

McEwan, E.H. (1975). [The adaptive significance of the growth patterns in cervids compared with other ungulate species.] *Zoologicheskii Zhurnal*, **54**, 1221–1232. (In Russian)

McEwan, E.H. & Whitehead, P.E. (1969). Changes in the blood constituents of caribou and reindeer occurring with age. *Canadian Journal of Zoology*, **47**, 557–562.

McEwan, E.H. & Whitehead, P.E. (1970). Seasonal changes in energy and nitrogen intake in reindeer and caribou. *Canadian Journal of Zoology*, **48**, 905–913.

McEwan, E.H. & Whitehead, P.E. (1972). Reproduction in female reindeer and caribou. *Canadian Journal of Zoology*, **50**, 43–46.

McEwan, E.H. & Whitehead, P.E. (1980). Plasma progesterone levels during anestrus, estrus and pregnancy in reindeer and caribou (*Rangifer tarandus*). In *Proceedings of 2nd International Reindeer/Caribou Symposium*, ed. E. Reimers, E. Gaare & S. Skjenneberg, pp. 324–328. Trondheim: Direktoratet for Vilt og Ferskvannfisk.

McEwan, E.H. & Wood, A.J. (1966). Growth and development of the barren-ground caribou. I. Heart girth, hindfoot length and body weight relationships. *Canadian Journal of Zoology*, **44**, 401–411.

Macpherson, A.H. (1965). The origin of diversity in mammals of the Canadian Arctic Tundra. *Systematic Zoology*, **14**, 153–173.

McRoberts, M.R., Hill, R. & Dalgarno, A.C. (1965). The effects of diets deficient in phosphorus and vitamin D, or calcium, on the skeleton and teeth of the growing sheep. Part 2. *Journal of Agricultural Science*, **65**, 11–14.

Mansell, W.D. (1971). Accessory corpora lutea in ovaries of white-tailed deer. *Journal of Wildlife Management*, **35**, 369–374.

Markgren, G. (1969). Reproduction of moose in Sweden. *Viltrevy*, **6**, 127–285.

Markgren, G. (1974). Factors affecting the reproduction of moose (*Alces alces*) in three different Swedish areas. *International Congress of Game Biology*, **9**, 67–70.

Marshall, F.H.A. (1937). On the change-over in the oestrous cycle in animals after transference across the equator, with further observations on the incidence of breeding seasons and the factors controlling sexual periodicity. *Proceedings of Royal Society*, Series B, **122**, 413–428.

Matthews, L.H. (1931). *South Georgia: the British Empire's sub-Antarctic Outpost*. Bristol: John Wright & Sons.

Ma Yi-ching, (1983). Status of reindeer in China. *Acta Zoologica Fennica*, **175**, 157–158.

Mech, L.D. (1970). *The Wolf: ecology and behaviour of an endangered species*. New York: Natural History Press.

Meldgaard, M. (1986). The Greenland caribou – zoogeography, taxonomy, and population dynamics. *Meddelelser om Grønland, Bioscience*, **20**, 1–88.

Meschaks, P. (1966). Sasongvariationer i testiklemorfologi hos ren (*Rangifer tarandus scandanavicus*). *Proceedings of Nordic Veterinary Congress*, **10**, 192–195.

Meschaks, P. & Nordkvist, M. (1962). Sexual cycle in the reindeer male. *Acta Veterinaria Scandanavica*, **3**, 151–162.

Messier, F. & Crête, M. (1984). Body condition and population regulation by food resources in moose. *Oecologia*, **65**, 44–50.

Messier, F. & Crête, M. (1985). Moose–wolf dynamics and the natural regulation of moose populations. *Oecologia*, **65**, 503–512.

Meurk, C.D. (1982). Regeneration of subantarctic plants on Campbell Island following exclusion of sheep. *New Zealand Journal of Ecology*, **5**, 51–58.

Michel, J.F. (1969). Arrested development of nematodes and some related phenomena. *Advances in Parasitology*, **12**, 280–366.

Michurin, L.N. (1967). [Reproduction of wild reindeer on the Taimyr Peninsula] *Zoologicheskii Zhurnal*, **46**, 1837–1841. (In Russian.)

Michurin, L.N. & Vakhtina, T.V. (1968). [Winter feeding of wild reindeer (*Rangifer tarandus* L.) on the arctic tundra of Taimyr.] *Zoologicheskii Zhurnal*, **47**, 477–479. (In Russian.)

Miller, D.R. (1976). Biology of the Kaminuriak population of barren-ground caribou. Part 3. Taiga winter range relationships and diet. *Canadian Wildlife Service, Report Series*, **36**, 1–41.

Miller, F.L. (1972). Eruption and attrition of mandibular teeth in barren-ground caribou. *Journal of Wildlife Management*, **36**, 606–612.

Miller, F.L. (1974). Biology of the Kaminuriak population of barren-ground caribou. Part 2. Dentition as an indicator of age and sex; compositon and socialisation of the population. *Canadian Wildlife Service, Report Series*, **31**, 1–88.

Miller, F.L. (1983). Restricted caribou harvest or welfare – northern native's dilemma. *Acta Zoologica Fennica*, **175**, 171–175.

Miller, F.L. (1986). Asymmetry in antlers of barren-ground caribou, Northwest Territories, Canada. *Rangifer*, Special Issue **1**, 195–202.

Miller, F.L. & Broughton, E. (1974). Calf mortality on the calving ground of Kaminuriak caribou. *Canadian Wildlife Service, Report Series*, **26**, 1–26.

Miller, F.L., Broughton, E. & Gunn, A. (1983). Mortality of newborn migratory barren-ground caribou calves, Northwest Territories, Canada. *Acta Zoolgica Fennica*, **175**, 155–156.

Miller, F.L., Cawley, A.J., Choquette, L.P.E. & Broughton, E. (1975). Radiographic examination of mandibular lesions in barren-ground caribou. *Journal of Wildlife Diseases*, **11**, 465–470.

Miller, F.L., Edmonds, E.J. & Gunn, A. (1982). Foraging behaviour of Peary caribou in response to springtime snow and ice conditions. *Canadian Wildlife Service, Occasional Paper*, **48**, 1–41.

Miller, F.L., Russell, R.H. & Gunn, A. (1975). The recent decline of Peary caribou on western Queen Elizabeth Islands of Arctic Canada. *Polarforschung*, **45**, 17–21.

Miller, F.L., Russell, R.H. & Gunn, A. (1977). Interisland movements of Peary caribou (*Rangifer tarandus pearyi*) on western Queen Elizabeth Islands, Arctic Canada. *Canadian Journal of Zoology*, **55**, 1029–1037.

Miller, F.L. & Tessier, G.D. (1971). Dental anomalies in caribou, *Rangifer tarandus*. *Journal of Mammalogy*, **52**, 164–174.

Mirachi, R.E., Scanlon, P.F. & Kirkpatrick, R.L. (1977). Annual changes in spermatozoan production and associated organs of white-tailed deer. *Journal of Wildlife Management*, **41**, 92–99.

Mitchell, B. (1973). The reproductive performance of wild Scottish red deer, *Cervus elaphus* L. *Journal of Reproduction and Fertility*, Supplement 19, 271–285.

Mitchell, B. & Brown, D. (1974). The effects of age and body size on fertility in female red deer (*Cervus elaphus* L.) *International Congress of Game Biology*, **9**, 89–98.

Mitchell, B., McCowan, D. & Nicholson, I.A. (1976). Annual cycles of body weight and condition in Scottish red deer. *Journal of Zoology, London*, **180**, 107–127.

Mitchell, B., McCowan, D. & Parrish, T. (1973). Some characteristics of natural mortality among wild Scottish red deer (*Cervus elaphus* L.). *Proceedings of International Congress of Game Biology*, **10**, 437–450.

Moen, A.N. (1973). *Wildlife Ecology*. San Francisco: Freeman & Co.

Moen, A.N. (1978). Seasonal changes in heart rates, activity, metabolism, and forage intake of white-tailed deer. *Journal of Wildlife Management*, **42**, 715–738.

Moore, D.M. (1968). The vascular flora of the Falkland Islands. *British Antarctic Survey, Scientific Reports*, **60**, 1–202.

Moors, P.J. & Atkinson, I.A.E. (1984). Predation on seabirds by introduced mammals, and factors affecting its severity. In *Status and Conservation of the World's Seabirds*, ed. J.P. Croxall, P.G.H. Evans & R.W. Schreiber, pp. 667–690. Cambridge: International Council for Bird Preservation.

Morrison, J.A. (1960). Ovarian characteristics in elk of known breeding history. *Journal of Wildlife Management*, **24**, 297–307.

Mossing, T. & Damber, J.E. (1981). Rutting behaviour and androgen variation in reindeer (*Rangifer tarandus* L.). *Journal of Chemical Ecology*, **7**, 377–389.

Muller-Schwarze, D. & Muller-Schwarze, C. (1983). Play behaviour in free-ranging caribou, *Rangifer tarandus*. *Acta Zoologica Fennica*, **175**, 121–124.

Murie, A. (1944). *The Wolves of Mt McKinley*. Washington, DC: Government Printing Office.

Naylor, L.L., Stern, R.O., Thomas, W.C. & Arobio, E.L. (1980). Socioeconomic evaluation of reindeer herding in Northwestern Alaska. *Arctic*, **33**, 246–272.

Neiland, K.A. (1970). Weight of dry marrow as an indicator of fat in caribou femurs. *Journal of Wildlife Management*, **34**, 904–907.

Newton, J.E. & Jackson, C. (1983). A note on the effect of dentition and age in sheep on the intake of herbage. *Animal Production*, **37**, 133–136.

Nieminen, M. (1980a). Evolution and taxonomy of the genus *Rangifer* in Northern Europe. In *Proceedings of 2nd International Reindeer/Caribou Symposium*, ed. E Reimers, E. Gaare & S. Skjenneberg, pp. 379–391. Trondheim: Direktoratet for Vilt og Ferskvannsfisk.

Nieminen, M. (1980b). Nutritional and seasonal effects on the haematology and blood chemistry in reindeer (*Rangifer tarandus tarandus* L.). *Comparative Biochemistry and Physiology*, **66A**, 399–413.

Nieminen, M. & Laitinen, M. (1986). Bone marrow and kidney fat as indicators of condition in reindeer. *Rangifer*, Special Issue **1**, 219–226.

Nikolaevskii, L.D. (1969). Diseases of reindeer. In *Reindeer Husbandry*, ed. P.S. Zhigunov, pp. 230–293. Jerusalem: Israel Program for Scientific Translations.

Nordkvist, M. (1967). Treatment experiments with systematic insecticides against the larvae of the reindeer grub fly (*Oedemagena tarandi* L.) and the reindeer nostril fly (*Cephanomyia trompe* L.). *Nordisk Veterinaermedicin*, **19** 281–293.

Nordkvist, M. (1971). Problems of veterinary medicine in reindeer breeding. *Veterinary Medical Review*, **2/3**, 405–414.

Nordkvist, M. (1980). Status of *Rangifer* in Sweden. In *Proceedings of 2nd International Reindeer/Caribou Symposium*, ed. by E. Reimers, E. Gaare & S. Skjenneberg, pp. 790–792. Trondheim: Direktoratet for Vilt og Ferskvannsfisk.

Nowasad, R F. (1973). Twinning in reindeer. *Journal of Mammalogy*, **54**, 781.

Nowasad, R.F. (1975). Reindeer survival in the Mackenzie Delta herd, birth to four months. *Biological Papers of University of Alaska, Special Report*, **1**, 199–208.

Olstad, O. (1930). Rats and reindeer in the Antarctic. *Scientific Results of the Norwegian Antarctic Expeditions*, **4**, 1–19.

Owen-Smith, N. (1981). The white rhino overpopulation problem and a proposed solution. In *Problems in Management of Locally Abundant Wild Mammals, ed. P.A. Jewell & S. Holt, pp. 129–150. New York: Academic Press.*

Packer, C. (1983). Sexual dimorphism: the horns of African antelopes. *Science*, **221**, 1191–1193.

Palmer, L.J. (1934). Raising reindeer in Alaska. *United States Department of Agriculture, Miscellaneous Publication*, **207**, 1–41.

Palmer L.J. & Rouse, C.H. (1945). Study of the Alaska tundra with reference to its reactions to reindeer and other grazing. *United States Department of Interior, Research Report*, **10**, 1–48.

Parker, G.R. (1972). Biology of the Kaminuriak population of barren-ground caribou. Part 1. Total numbers, mortality, recruitment and seasonal distribution. *Canadian Wildlife Service, Report Series*, **20**, 1–93.

Parker, G.R. (1973). Distribution and densities of wolves within barren-ground caribou range in northern mainland Canada. *Journal of Mammalogy*, **54**, 341–348.

Parker, G.R. (1975). An investigation of caribou range on Southampton Island, NWT. *Canadian Wildlife Service, Report Series*, **33**, 1–82.

Parker, G.R. (1981). Physical and reproductive characteristics of an expanding woodland caribou population (*Rangifer tarandus caribou*) in northern Labrador. *Canadian Journal of Zoology*, **59**, 1929–1940.

Parker, G.R. & Luttich, S. (1986). Characteristics of the wolf (*Canis lupus labradorius* Goldman) in Northern Quebec and Labrador. *Arctic*, **39**, 145–149.

Parovshchikov, V.Y. (1965). [Distribution and number of wild reindeer of Archangelsk North.] *Zoologischeskii Zhurnal*, **44**, 276–283. (In Russian.)

Pascal, M. (1980). Structure et dynamique de la population de chats harets de l'archipel Kerguelen. *Mammalia*, **44**, 161–182.

Pascal, M. (1982). Les espèces mammaliennes introduits dans L'Archipel des Kerguelen. *Comité National Français des Recherches Antarctiques*, **51**, 269–280.

Payne, M.R. (1977). Growth of a fur seal population. *Philosophical Transactions of Royal Society of London*, Series B, **279**, 612–619.

Peek, J.M. (1980). Natural regulation of ungulates (what constitutes a wilderness?) *Wildlife Society Bulletin*, **8**, 217–227.

Pegau, R.E. (1970a). Effect of reindeer trampling and grazing on lichens. *Journal of Range Management*, **23**, 95–97.

Pegau, R.E. (1970b). Succession in two exclosures near Unalakleet, Alaska. *Canadian Field-Naturalist*, **84**, 175–177.

Pegau, R.E. (1975). Analysis of the Nelchina caribou range. *Biological Papers of University of Alaska, Special Report*, **1**, 316–323.

Pellew, R.A. (1983). The giraffe and its food resource in the Serengeti. II. Response of the giraffe population to changes in the food supply. *African Journal of Ecology*, **21**, 269–283.

Peterson, R.L. (1955). *North American Moose*. Toronto: Toronto University Press.

Peterson, R.O. (1977). *Wolf Ecology and Prey Relationships on Isle Royale*. National Park Service Scientific Monograph Service No. 11. Washington: US National Park Service.

Pimlott, D.H. (1967). Wolf predation and ungulate populations. *American Zoologist*, **7**, 267–278.

Pollock, A.M. (1974). Seasonal changes in appetite and sexual condition in red deer stags on a six-month photoperiod. *Journal of Physiology, London*, **224**, 95–96P.

Poole, R.W. (1964). *An Introduction to Quantitative Ecology*. New York: McGraw-Hill.

Pratt, R.M. & Lewis Smith, R.I. (1982). Seasonal trends in the chemical composition of reindeer forage plants on South Georgia. *Polar Biology*, **1**, 13–32.

Preobrazhenskii, B.V. (1969). Management and breeding of reindeer. In *Reindeer Husbandry*, ed. P.S. Zhigunov, pp. 78–128. Jerusalem: Israel Program for Scientific Translations.

Prior, R. (1968). *The Roe Deer of Cranborne Chase: an ecological survey*. London: Oxford University Press.

Prowse, D.L., Trilling, J.S. & Luick, J.R. (1980). Effects of antler removal on mating behaviour of reindeer. In *Proceedings of 2nd International Reindeer/Caribou Symposium*, ed. E. Reimers, E. Gaare & S. Skjenneberg, pp. 528–536. Trondheim: Direktoratet for Vilt og Ferskvannfisk.

Pruitt, W.O. (1959). Snow as a factor in the winter ecology of the barren-ground caribou (*Rangifer arcticus*). *Arctic*, **12**, 159–179.

Pruitt, W.O. (1960a). Behaviour of the barren-ground caribou. *Biological Papers of University of Alaska*, **3**, 1–43.

Pruitt, W.O. (1960b). Animals in the snow. *Scientific American*, **202**, 60–68.

Pruitt, W.O. (1966). Function of the brow tine in caribou antlers. *Arctic*, **19**, 111–113.

Pryadko, E.I. (1976). [*Helminths of Deer*]. Alma-Ata: Icdateistvo Nauka Kazakhskoi. (In Russian.)

Pulliainen, E. (1980). Status of *Rangifer* in the Karelian ASSR. In *Proceedings of 2nd International Reindeer/Caribou Symposium*, ed. by E. Reimers, E. Gaare & S. Skjenneberg, pp. 771–773. Trondheim: Direktoratet for Vilt og Ferskvannsfisk.

Pulliainen, E. & Siivonen, L. (1980). Status of *Rangifer* in Finland. In *Proceedings of 2nd International Reindeer/Caribou Symposium*, ed. E. Reimers, E. Gaare & S. Skjenneberg, pp. 760–763. Trondheim: Direktoratet for Vilt og Ferskvannsfisk.

Punsvik, T., Syvertsen, A. & Staaland, H. (1980). Reindeer grazing in Adventdalen, Svalbard. In *Proceedings of 2nd International Reindeer/*

Caribou Symposium, ed. E. Reimers, E. Gaare & S. Skjenneberg, pp. 115–123. Trondheim: Direktoratet for Vilt og Ferskvannsfisk.

Pye, T. (1984). Biology of the house mouse (*Mus musculus*) on Macquarie Island. *Tasmanian Naturalist*, **79**, 6–10.

Pye, T. & Bonner, W.N. (1980). Feral brown rats, *Rattus norvegicus*, in South Georgia (South Atlantic Ocean). *Journal of Zoology, London*, **192**, 237–255.

Ralls, K., Brugger, K. & Ballou, J. (1979). Inbreeding and juvenile mortality in small populations of ungulates. *Science*, **206**, 1101–1103.

Ransom, A.B. (1966). Breeding seasons of white-tailed deer in Manitoba. *Canadian Journal of Zoology*, **44**, 59–62.

Rasmussen, D.I. (1941). Biotic communities of Kaibab Plateau, Arizona. *Ecological Monographs*, **11**, 229–275.

Razmakhnin, V.Y. (1986). Wild reindeer in the USSR, their protection and utilization. *Rangifer*, Special Issue **1**, 347–349.

Rehbinder, C. & Christennson, D. (1977). The presence of gastro-intestinal parasites in autumn-slaughtered reindeer bulls. *Nordisk Veterinaermedicin*, **29**, 556–557.

Rehbinder, C., Edqvist, L.-E., Riesten-Århed, U. & Nordkvist, M. (1981). Progesterone in pregnant and non-pregnant reindeer. *Acta Veterinaria Scandanavica*, **22**, 355–359.

Reimers, E. (1972). Growth in domestic and wild reindeer in Norway. *Journal of Wildlife Management*, **36**, 612–619.

Reimers, E. (1975). Age and sex structure in a hunted population of reindeer in Norway. *Biological Papers of University of Alaska, Special Report*, **1**, 181–188.

Reimers, E. (1977). Population dynamics in two subpopulations of reindeer in Svalbard. *Arctic and Alpine Research*, **9**, 369–381.

Reimers, E. (1980). Activity pattern: the major determinant for growth and fattening in *Rangifer*. In *Proceedings of 2nd International Reindeer/Caribou Symposium*, ed. E. Reimers, E. Gaare & S. Skjenneberg, pp. 466–474. Trondheim: Direktoratet for Vilt og Ferskvannsfisk.

Reimers, E. (1982). Winter mortality and population trends of reindeer on Svalbard, Norway. *Arctic and Alpine Research*, **14**, 295–300.

Reimers, E. (1983a). Mortality in Svalbard reindeer. *Holarctic Ecology*, **6**, 141–149.

Reimers, E. (1983b). Reproduction in wild reindeer in Norway. *Canadian Journal of Zoology*, **61**, 211–217.

Reimers, E. (1983c). Growth rate and body size differences in *Rangifer*, a study of causes and effects. *Rangifer*, **3**, 3–15.

Reimers, E., Klein, D.R. & Sørumgard, R. (1983). Calving time, growth rate and body size of Norwegian reindeer on different ranges. *Arctic and Alpine Research*, **15**, 107–108.

Reimers, E. & Ringberg, T. (1983). Seasonal changes in body weights of Svalbard reindeer from birth to maturity. *Acta Zoologica Fennica*, **175**, 69–72.

Reimers, E., Ringberg, T. & Sørumgard, R. (1982). Body composition of Svalbard reindeer. *Canadian Journal of Zoology*, **60**, 1812–1821.

Reimers, E., Villmo, L., Gaare, E. & Skogland, T. (1980). Status of *Rangifer* in Norway including Svalbard. In *Proceedings of 2nd International Reindeer/ Caribou Symposium*, ed. E. Reimers, E. Gaare & S. Skjenneberg, pp. 774–785. Trondheim: Direktoratet for Vilt og Ferskvannsfisk.

Richards, F.J. (1959). A flexible growth function for empirical use. *Journal of Experimental Botany*, **10**, 290–300.

Richardson, C., Richards, M., Terlecki, S. & Miller, W.M. (1979). Jaws of adult culled ewes. *Journal of Agricultural Science*, **93**, 521–529.

Ricklefs, R.E. (1973). *Ecology*. London: Nelson.

Rieck, W. (1955). [The birth of roe, red and fallow deer in central Europe.] *Zeitschrift für Jagdwissenschaft*, **1**, 69–75. (In German.)

Riney, T. (1955). Evaluation of condition of free-ranging red deer with reference to New Zealand. *New Zealand Journal of Science and Technology*, B, **46**, 429–463.

Riney, T. (1964). The impact of introductions of large herbivores on the tropical environment. *International Union for the Conservation of Nature Publications, New Series*, **4**, 261–273.

Ringberg Lund-Larsen, T. (1977). Relation between testosterone levels in serum and proteolytic activity in neck muscles of the Norwegian reindeer (*Rangifer tarandus tarandus*). *Acta Zoologica, Stockholm*, **58**, 61–63.

Ringberg, T. (1979). The Spitzbergen reindeer – winter dormant ungulate? *Acta Physiologica Scandanavica*, **105**, 268–273.

Ringberg, T. & Aakvaag, A. (1982). The diagnosis of early pregnancy and missed abortion in European and Svalbard reindeer (*Rangifer tarandus* and *Rangifer tarandus platyrhyncus*). *Rangifer*, **2**, 26–30.

Ringberg, T., Jacobsen, E., Ryg, M. & Krog, J. (1978). Seasonal changes in level of growth hormone, somatomedin and thyroxine in free-ranging, semi-domesticated Norwegian reindeer (*Rangifer tarandus tarandus* L.) *Comparative Biochemistry and Physiology*, **60A**, 123–126.

Robbins, C.T. & Robbins, B.L. (1979). Fetal and neonatal growth patterns and maternal reproductive effort in ungulates and subungulates. *American Naturalist*, **114**, 101–116.

Robinette, W.L., Gashwiler, J.S., Low, J.B. & Jones, D.A. (1957). Differential mortality by sex and age amongst mule deer. *Journal of Wildlife Management*, **21**, 1–16.

Roby, D.D., Thing, H. & Brink, K.L. (1984). History, status and taxonomic identity of caribou (*Rangifer tarandus*) in Northwest Greenland. *Arctic*, **37**, 23–30.

Roine, K. (1970). Biometric observations of the genital organs of the reindeer cow. *Proceedings of Nordic Veterinary Congress*, **11**, 262.

Roneus, O. & Nordkvist, M. (1962). Cerebrospinal and muscular nematodiasis (*Elaphostrongylus rangiferi*) in Swedish reindeer. *Acta Veterinaria Scandanavica*, **3**, 201–225.

Rosing, J. (1956). Experimental reindeer husbandry in West Greenland. *Polar Record*, **8**, 272–273.

Rounsevell, D.E. & Brothers, N.P. (1984). The status and conservation of seabirds at Macquarie Island. In *Status and Conservation of the World's Seabirds*, ed. J.P. Croxall, P.G.H. Evans & R.W. Schreiber, pp. 587–592. Cambridge: International Council for Bird Preservation.

Rudge, M.R. (1970). Dental and periodontal abnormalities in two populations of feral goats (*Capra hircus* L.) in New Zealand. *New Zealand Journal of Science*, **13**, 260–267.

Rudge, M.R. & Campbell, D.J. (1977). The history and present status of goats on the Auckland Islands (New Zealand subantarctic) in relation to vegetation changes induced by man. *New Zealand Journal of Botany*, **15**, 221–254.

Ryg, M. (1986). Physiological control of growth, reproduction and lactation in deer. *Rangifer*, Special Issue **1**, 261–266.

Ryg, M. & Jacobsen, E. (1982a). Seasonal changes in growth rate, feed intake, growth hormone, and thyroid hormones in young male reindeer (*Rangifer tarandus tarandus*). *Canadian Journal of Zoology*, **60**, 15–23.

Ryg, M. & Jacobsen, E. (1982b). Effects of castration on growth and food intake cycles in young male reindeer (*Rangifer tarandus tarandus*). *Canadian Journal of Zoology*, **60**, 942–945.

Ryg, M. & Jacobsen, E. (1982c). Effects of thyroid hormones and prolactin on food intake and weight changes in young male reindeer (*Rangifer tarandus tarandus*). *Canadian Journal of Zoology*, **60**, 1562–1567.

Røed, K. H. (1985). Comparison of the genetic variation in Svalbard and Norwegian reindeer. *Canadian Journal of Zoology*, **63**, 2035–2042.

Scheffer, V.B. (1951). The rise and fall of a reindeer herd. *Scientific Monthly*, **73**, 356–362.

Scheffer, V.B. (1955). Body size with relation to population density in mammals. *Journal of Mammalogy*, **36**, 493–515.

Schlanger, S.O. & Gillett, G.W. (1976). A geological perspective of the upland biota of Layson atoll (Hawaiian Islands). *Biological Journal of Linnaean Society*, **8**, 205–216.

Schmitt, E.V. (1936). [A determination of the period of gestation of domestic reindeer]. *Sovetskoye olenevodstvo*, **8**, 35–43. (In Russian.)

Scott, J.J. (1983). Landslip revegetation and rabbits, Subantarctic Macquarie Island. *Proceedings of Ecological Society of Australia*, **12**, 170–171.

Scotter, G.W. (1967). The winter diet of caribou in northern Canada. *Canadian Field-Naturalist*, **81**, 33–39.

Scotter, G.W. (1972). Reindeer ranching in Canada. *Journal of Range Management*, **25**, 167–174.

Segonzac, M. (1972). Données recentes sur la faune des iles Saint-Paul et Nouvelle Amsterdam. *L'oiseau et la revue français d'ornithologie*, **42**, 3–68.

Selkirk, P.M., Costin, A.B., Seppelt, R.D. & Scott, J.J. (1983). Rabbits, vegetation and erosion on Macquarie Island. *Proceedings of Linnaean Society of New South Wales*, **106**, 337–346.

Semenov-Tian-Shanskii, O.I. (1975). The status of wild reindeer in the USSR, especially the Kola Peninsula. *Biological Papers of University of Alaska, Special Report*, **1**, 155–161.

Sengbusch, H.G. (1977). Review of oribatid mitc–anoplocephalan tapeworm relationships (Acari; Oribatei; Cestoda; Anoplocephalidae). In *Biology of Oribatid Mites*, ed. D.L. Dindal, pp. 87–102. Syracuse: State University of New York.

Shalaeva, N.M. (1975). The helminth fauna of the wild reindeer of western Taimyr. *Biological Papers of University of Alaska, Special Report*, **1**, 507–508.

Shank, C.C. Wilkinson, P.F. and Penner, D.F. (1978). Diet of Peary caribou, Banks Island, NWT. *Arctic*, **31**, 125–132.

Sharp, H.S. (1977). The caribou-eater Chipewyan: bilaterality, strategies of caribou hunting, and the fur trade. *Arctic Anthropology*, **14**, 35–40.

Shope, R.E., MacNamara, L.G. & Mangold, R. (1960). A virus-induced epizootic hemorrhagic disease of the Virginia white-tailed deer (*Odocoileus virginianus*). *Journal of Experimental Medicine*, **111**, 155–170.

Short, H.L. (1971). Forage digestibility and diet of deer on southern upland range. *Journal of Wildlife Management*, **35**, 698–706.

Short, R.V. (1980). Sexual selection: the meeting point of endocrinology and sociobiology. *Proceedings of 6th International Congress of Endocrinology*, ed. I.A. Cumming, J.W. Funder & F.A.O. Mendelsohn, pp. 49–57. Canberra: Australian Academy of Science.

Short, R.V. & Mann, T.R.R. (1966). The sexual cycle of a seasonally breeding mammal, the roebuck (*Capreolus capreolus* L.). *Journal of Reproduction and Fertility*, **12**, 337–351.

Siegel, S. (1956). *Non-parametric Statistics*. New York: McGraw-Hill.

Sieveking, A. (1979). *The Cave Artists*. London: Thames & Hudson.

Silver, H., Colovus, N.F., Holter, J.B. & Hayes, H.H. (1969). Fasting metabolism of white-tailed deer. *Journal of Wildlife Management*, **33**, 490–498.

Simkin, D.W. (1965). Reproduction and productivity of moose in northwestern Ontario. *Journal of Wildlife Management*, **29**, 740–750.

Sinclair, A.R.E. (1975). The resource limitation of trophic levels in tropical grassland ecosystems. *Journal of Animal Ecology*, **44**, 497–520.

Sinclair, A.R.E. (1977). *The African Buffalo: A study of resource limitation*. Chicago: Chicago University Press.

Sinclair, A.R.E. (1979). The eruption of the ruminants. In *Serengeti: Dynamics of an ecosystem*, ed. A.R.E. Sinclair & M. Norton-Griffiths, pp. 82–103. Chicago: Chicago University Press.

Sinclair, A.R.E. (1985). Does interspecific competition or predation shape the African ungulate community? *Journal of Animal Ecology*, **54**, 899–918.

Skira, I.J. (1978). Reproduction of the rabbit, *Oryctolagus cuniculus* (L.), on Macquarie Island, Subantarctic. *Australian Wildlife Research*, **5**, 317–326.

Skjenneberg, S. (1983). Advance in the study of veterinary medicine in the reindeer/caribou since the year 1978. *Acta Zoologica Fennica*, **175**, 91–98.

Skjenneberg, S. & Slagsvold, L. (1968). *Reindriften og dens naturgrunnlag*. Oslo: Universitetsforlaget.

Skogland, T. (1975). Wild reindeer range use and selectivity in South Norway. *Biological Papers of University of Alaska, Special Report*, **1**, 342–354.

Skogland, T. (1978). Characterisitics of the snow cover and its relationship to wild mountain reindeer (*Rangifer tarandus tarandus* L.) feeding strategies. *Arctic and Alpine Research*, **10**, 569–580.

Skogland, T. (1980). Comparative summer feeding strategies of arctic and alpine *Rangifer*, *Journal of Animal Ecology*, **49**, 81–98.

Skogland, T. (1983). The effects of density-dependent resource limitation on size of wild reindeer. *Oecologia (Berlin)*, **60**, 156–168.

Skogland, T. (1984). Wild reindeer foraging-niche organisation. *Holarctic Ecology*, **7**, 345–379.

Skogland, T. (1985). The effects of density-dependent resource limitations on the demography of wild reindeer. *Journal of Animal Ecology*, **54**, 359–374.

Skoog, R.O. (1968). *Ecology of the caribou (Rangifer tarandus granti) in Alaska*. Ph.D. thesis, University of California, Berkeley.

Skrjabin, K.I., Shikhobalova, N.P., Sobolov, A.A., Paramonov, A.A. & Sudarikov, V.E. (1954). [*Descriptive catalogue of parasitic nematodes. Volume 4.*] Moscow: Soviet Academy of Science. (In Russian, host lists under Latin names.)

Skrobov, V.D. (1984). Human intevention and wild reindeer. In *Wild Reindeer of the Soviet Union*, ed. E.E. Syroechkovskii, pp. 90–94. Washington, DC: United States Department of the Interior.

Skuncke, F. (1969). Reindeer ecology and management in Sweden. *Biological Papers of University of Alaska*, **8**, 1–81.

Snedecor, G.W. & Cochran, W.G. (1967). *Statistical Methods*, 6th edn. Iowa: Iowa State University Press.

Sokal, R.R. & Rohlf, F.J. (1969). *Biometry*. San Francisco: Freeman & Co.

Sokolov, I.I. (1959). [*Fauna of the USSR: Mammals*, Vol. 1 (3), *Ungulates*]. Moscow: Academy of Sciences. (In Russian.)

Spiess, A.E. (1979). *Reindeer and Caribou Hunters*. New York: Academic Press.

Staaland, H., Jacobsen, E. & White, R.G. (1979). Comparison of the digestive tract in Svalbard and Norwegian reindeer. *Arctic and Alpine Research*, **11**, 457–466.

Stager, J.K. (1984). Reindeer herding as private enterprise in Canada. *Polar Record*, **22**, 127–136.

Staines, B.W., Crisp, J.M. & Parrish, T. (1982). Differences in the quality of food eaten by red deer (*Cervus elaphus*) stags and hinds in winter. *Journal of Applied Ecology*, **19**, 65–77.

Stoddart, D.R. (1981). History of goats in the Aldabra archipelago. *Atoll Research Bulletin*, **255**, 23–26.

Stokkan, K.-A., Hove, K. & Carr, W.R. (1980). Plasma concentrations of testosterone and luteinising hormone in rutting reindeer bulls (*Rangifer tarandus*). *Canadian Journal of Zoology*, **58**, 2081–2083.

Strandgaard, H. (1972a). An investigation of corpora lutea, embryonic development, and time of birth of roe deer (*Capreolus capreolus*) in Denmark. *Danish Review of Game Biology*, **6**, 3–22.

Strandgaard, H. (1972b). The roe deer (*Capreolus capreolus*) population at Kalo and the factors regulating its size. *Danish Review of Game Biology*, **7**, 1–205.

Strandgaard, H., Holthe, V., Lassen, P. & Thing, H. (1983). *Rensdyrundersøgelser i Vestgrønland 1977–1982*. Kalø: Vildbiologisk Station. (In Danish.)

Strange, I.J. (1981). *The Falkland Islands*, 2nd edn. Newton Abbot: David & Charles.

Sugden, D.E. & Clapperton, C.M. (1977). The maximum ice extent of island groups in the Scotia Sea, Antarctica. *Quaternary Research*, **7**, 268–282.

Sykes, A.R. (1978). The effect of subclinical parasitism in sheep. *Veterinary Record*, **102**, 32–34.

Syrjälä-Qvist, L. & Salonen, J. (1983). Effect of protein and energy supply on nitrogen utilisation in reindeer. *Acta Zoologica Fennica*, **175**, 53–55.

Syroechkovskii, E.E. (1984). Overview of the problem of wild reindeer in Soviet Union. In *Wild Reindeer of the Soviet Union*, ed. E.E. Syroechkovskii, pp. 6–44. Washington, DC: United States Department of the Interior.

Taber, R.D. & Dasmann, R.F. (1957). The dynamics of three natural populations of the deer *Odocoileus hemionius columbianus*. *Ecology*, **38**, 233–246.

Tandler, J. & Grosz, S. (1913). *Die biologischen Grundlagen der sekündären Geschlechts charaktere*. Berlin: Springer Verlag. Taylor, R.H. (1971). Influence of man on vegetation and wildlife of Enderby and Rose Islands, Auckland Islands. *New Zealand Journal of Botany*, **9**, 225–268.

Taylor, R.H. (1979). How the Macquarie Island parakeet became extinct. *New Zealand Journal of Ecology*, **2**, 42–45.

Tener, J.S. (1965). *Muskoxen in Canada*. Ottawa: Canadian Wildlife Service.

Thing, H. (1977). Behavior, mechanics and energetics associated with winter cratering by caribou in northwestern Alaska. *Biological Papers of University of Alaska*, **18**, 1–41.

Thing, H. (1980). Status of *Rangifer* in Greenland. In *Proceedings of 2nd International Reindeer/Caribou Symposium*, ed. E. Reimers, E. Gaare & S. Skjenneberg, pp. 764–765. Trondheim: Direktoratet for Vilt og Ferksvannsfisk.

Thing, H. (1981). *Feeding ecology of the west Greenland caribou (Rangifer tarandus groenlandicus Gmlin.) in the Holsteinborg-Sdr. Strømfjord region*. Ph.D. thesis, University of Aarhus, Denmark.

Thing, H. (1984). Feeding ecology of the west Greenland caribou (*Rangifer tarandus groenlandicus*) in the Sisimiut–Kangerlussuaq Region. *Danish Review of Game Biology*, **12**(3), 1–53.

Thing, H. & Clausen, B. (1980). Summer mortality among caribou calves in Greenland. In *Proceedings of 2nd International Reindeer/Caribou Symposium*, ed. E. Reimers, E. Gaare & S. Skjenneberg, pp. 434–437. Trondheim: Direktoratet for Vilt og Ferskvannsfisk.

Thing, H., Olesen, C.R. & Aastrup, P. (1986). Antler possession in west Greenland female caribou in relation to population characteristics. *Rangifer*, Special Issue **1**, 297–304.

Thomas, D.C. (1970). *The ovary, reproduction and productivity of female Columbian black-tailed deer*. Ph.D. thesis, University of British Columbia, Vancouver.

Thomas, D.C. (1982). The relationship between fertility and fat reserves of Peary caribou. *Canadian Journal of Zoology*, **60**, 597–602.

Thomas, D.C. & Edmonds, J. (1983). Rumen contents and habitat selection of Peary caribou in winter, Canadian Arctic archipelago. *Arctic and Alpine Research*, **15**, 97–105.

Thomas, D.C. & Hervieux, D.P. (1986). The late winter diets of barren-ground caribou in North-Central Canada. *Rangifer*, Special Issue **1**, 305–310.

Thomas, W.C. & Arobio, E.L. (1983). Public policy: implications for Alaska reindeer herd management. *Acta Zoologica Fennica*, **175**, 177–179.

Thórisson, S. (1983). [*The history of reindeer in Iceland and reindeer study 1979–1981.*] Reykjavik: Orkustofnun. (In Icelandic.)

Timisjärvi, J., Nieminen, M. & Saari, E. (1981). Haematological values in reindeer. *Journal of Wildlife Management*, **45**, 976–981.

Timisjärvi, J., Nieminen, M., Roine, K., Koskinen, M. & Laaksonen, H. (1982). Growth in the reindeer. *Acta Veterinaria Scandanavica*, **23**, 603–618.

Treude, E. (1968). The development of reindeer husbandry in Canada. *Polar Record*, **144**, 15–19.

Trivers, R.L. (1972). Parental investment and sexual selection. In *Sexual Selection and the Descent of Man 1871–1971*, ed. B. Campbell, pp. 136–179. Chicago: Aldine-Atherton.

Trudell, J. & White, R.G. (1981). The effect of forage structure and availability on food intake, biting rate, bite size and daily eating time of reindeer. *Journal of Applied Ecology*, **18**, 63–81.

Tyler, N.J.C. (1986). The relationship between the fat content of Svalbard reindeer in autumn and their death from starvation in winter. *Rangifer*, Special Issue 1, 311–314.

Tyler, N.J.C. (1987a). *Natural limitation of the abundance of the High Arctic Svalbard reindeer*. Ph.D. thesis, University of Cambridge.

Tyler, N.J.C. (1987b). Body composition and energy balance of pregnant and non-pregnant Svalbard reindeer during winter. *Symposium of Zoological Society of London*, 57, 203–229.

Utsi, M.N.P. (1948). The reindeer-breeding methods of northern Lapps. *Man*, 114, 1–5.

van Aarde, R.J. (1978). Reproduction and population ecology in the feral house cat, *Felis catus*, at Marion Island. *Carnivore Genetics Newsletter*, 3, 288–313.

van Aarde, R.J. (1979). Distribution and density of the feral house cat, *Felis catus*, at Marion Island. *South African Journal of Antarctic Research*, 9, 14–19.

van Aarde, R.J. (1980). The diet and feeding behaviour of feral cats, *Felis catus*, at Marion Island. *South African Journal of Wildlife Research*, 10, 123–128.

van Aarde, R.J. & Robinson, T.J. (1980). Gene frequencies in feral cats on Marion Island, South Africa. *Journal of Heredity*, 71, 366–368.

van Aarde, R.J. & Skinner, J.D. (1982). The feral cat population at Marion Island: characteristics, colonisation and control. *Comité National Français des Recherches Antarctiques*, 51, 281–288.

Van Ballenberghe, V. (1985). Wolf predation on caribou: the Nelchina herd case history. *Journal of Wildlife Management*, 49, 711–720.

van Rensburg, P.J.J. (1985). The feeding ecology of a decreasing feral house cat, *Felis catus*, population at Marion Island. In *Antarctic Nutrient Cycles and Food Webs*, ed. W.R. Siegfried, P.R. Condy & R.M. Laws, pp. 620–624. Berlin: Springer Verlag.

Veitch, C.R. (1985). Methods of eradicating feral cats from offshore islands in New Zealand. In *Conservation of Island Birds*, ed. P.J. Moors, pp. 125–141. Cambridge: International Council for Bird Preservation.

Verme, L.J. (1965). Reproduction studies on penned white-tailed deer. *Journal of Wildlife Management*, 29, 74–79.

Verme, L.J. (1983). Sex ratio variation in *Odocoileus*: a critical review. *Journal of Wildlife Management*, 47, 573–582.

Vibe, C. (1967). Arctic animals in relation to climatic fluctuations. *Meddelelser om Grønland*, 170, 1–227.

Vogel, M. (1985). The distribution and ecology of epigeic invertebrate fauna on the subantarctic island of South Georgia. *Spixiana*, 8, 153–163.

Vogel, M., Remmert, H. & Lewis Smith, R.I. (1984). Introduced reindeer and their effects on the vegetation and the epigeic invertebrate fauna of South Georgia (subantarctic). *Oecologia (Berlin)*, 62, 102–109.

de Vos, A., Manville, R.H. & Van Gelder, R.G. (1956). Introduced mammals and their influence on native biota. *Zoologica, New York*, 41, 163–194.

Wace, N.M. & Holdgate, M.W. (1976). *Man and Nature in the Tristan da Cuhna Islands*. IUCN Monograph No. 6. Morges: International Union for the Conservation of Nature.

Wales, R.A., Milligan, L.P. & McEwan, E.H. (1975). Urea recycling in caribou, cattle and sheep. *Biological Papers of University of Alaska, Special Report*, 1, 297–307.

Wallmo, O.C. (ed.) (1981). *Mule and Black-tailed Deer of North America*. Lincoln: University of Nebraska Press.

Walton, D.W.H. (1975). European weeds and other alien species in the Subantarctic. *Weed Research*, **15**, 271–282.

Walton, D.W.H. (1982). Floral phenology in the South Georgian vascular flora. *British Antarctic Survey Bulletin*, **55**, 11–25.

Walton, D.W.H. (1984). The terrestrial environment. In *Antarctic Ecology*, ed. R.M. Laws, pp. 1–60. London: Academic Press.

Walton, D.W.H. (1985). The Sub-Antarctic Islands. In *Key Environments – Antarctica*, ed. W.N. Bonnner & D.W.H. Walton, pp. 293–317. Oxford: Pergamon Press.

Walton, D.W.H. & Lewis Smith, R.I. (1980). The chemical composition of South Georgian vegetation. *British Antarctic Survey Bulletin*, **49**, 117–135.

Watson, G.E. (1975). *Birds of the Antarctic and Sub-Antarctic*. Washington: American Geophysical Union.

Watson, R.M. (1969). Reproduction of wildebeest, *Connochaetes taurinus albojubatus* Thomas, and its significance to conservation. *Journal of Reproduction and Fertility, Supplement*, **6**, 287–310.

West, N.O. & Nordan, H.C. (1976). Hormonal regulation of reproduction and the antler cycle in the male Columbian black-tailed deer (*Odocoileus hemionus columbianus*). Part I. Seasonal changes in the histology of the reproductive organs, serum testosterone, sperm production and the antler cycle. *Canadian Journal of Zoology*, **54**, 1617–1626.

White, R.G. (1983). Foraging patterns and their multiplier effects on productivity of northern ungulates. *Oikos*, **40**, 377–384.

White, R.G., Bunnell, F.L., Gaare, E., Skogland, T. & Hubert, T. (1981). Ungulates on Arctic ranges. In *Tundra Ecosystems: A comparative analysis*, ed. L.C. Bliss, J.B. Cragg, D.W. Heal & J.J. Moore, pp. 397–483. Cambridge: Cambridge University Press.

White, R.G., Thomson, B.R., Skogland, T., Person, S.J., Russell, D.E., Holleman, D.F. & Luick, J.R. (1975). Ecology of caribou at Prudhoe Bay, Alaska. *Biological Papers of University of Alaska, Special Report*, **2**, 151–201.

White, R.G. & Trudell, J. (1980). Habitat preference and forage consumption by reindeer and caribou near Atkasook, Alaska. *Arctic and Alpine Research*, **12**, 511–529.

White, R.W. (1961). Some foods of the white-tailed deer in Arizona. *Journal of Wildlife Management*, **25**, 404–409.

Whitehead, G.K. (1972). *Deer of the World*. London: Constable.

Whitehead, P.E. & McEwan, E.H. (1973). Seasonal variation in the plasma testosterone concentration of reindeer and caribou. *Canadian Journal of Zoology*, **51**, 651–658.

Whitehead, P.E. & West, N.O. (1977). Metabolic clearance and production rates of testosterone at different times of year in male caribou and reindeer. *Canadian Journal of Zoology*, **55**, 1692–1697.

Whitten, K.R. & Cameron, R.D. (1980). Nutrient dynamics of caribou forage on Alaska's Arctic slope. In *Proceedings of 2nd International Reindeer/ Caribou Symposium*, ed. E. Reimers, E. Gaare & S. Skjenneberg, pp. 159–166. Trondheim: Direktoratet for Vilt og Ferskvannsfisk.

Wika, M. (1980). On growth of reindeer antlers. In *Proceedings of 2nd International Reindeer/Caribou Symposium*, ed. E. Reimers, E. Gaare & S. Skjenneberg, pp. 416–421. Trondheim: Direktoratet for Vilt og Ferskvannsfisk.

Wika, M. (1982). Foetal stages of antler development. *Acta Zoologica, Stockholm*, **63**, 187–189.

Wilkins, G.H. (1925). Gough Island. *Journal of Botany, London*, **63**, 65–70.

Williams, A.J. (1984). The status and conservation of seabirds on some islands in the African sector of the Southern Ocean. In *Status and Conservation of the World's Seabirds*, ed. J.P. Croxall, P.G.H. Evans & R.W. Schreiber, pp. 627–635. Cambridge: International Council for Bird Preservation.

Williams, G.C. (1975). *Sex and Evolution*. Princeton: Princeton University Press.

Williams, T.M. & Heard, D.C. (1986). World status of wild *Rangifer tarandus* populations. *Rangifer*, Special Issue **1**, 19–28.

Williamson, M. (1981). *Island Populations*. Oxford: Oxford University Press.

Wilson, P.R. & Orwin, D.F.G. (1964). The sheep population of Campbell Island. *New Zealand Journal of Science*, **7**, 460–490.

Wislocki, G.B. (1943). Studies on the growth of deer antlers. II. Seasonal changes in the male reproductive tract of the Virginia deer (*Odocoileus virginianus borealis*) with a discussion of the factors controlling the antler-gonad periodicity. In *Essays in Biology in Honor of Herbert M. Evans*, pp. 631–653. Berkeley: University of California Press.

Wislocki, G.B., Aub, J.C. & Waldo, C.M. (1947) The effects of gonadectomy and the administration of testosterone proprionate on the growth of antlers in male and female deer. *Endocrinology*, **40**, 202–224.

Wodzicki, K. (1950). Introduced mammals of New Zealand. *Department of Scientific and Industrial Research, Bulletin*, **98**, 1–255.

Wodzicki, K. (1961). Ecology and management of introduced ungulates in New Zealand. *La Terre et la Vie*, **1**, 130–157.

Wood, A.J., Cowan, I.McT. & Nordan, H.C. (1962). Periodicity of growth in ungulates as shown by deer the genus *Odocoileus*. *Canadian Journal of Zoology*, **40**, 593–603.

Yousef, M.K., Cameron, R.D. & Luick, J.R. (1971). Seasonal changes in hydrocortisone secretion rate of reindeer, *Rangifer tarandus*. *Comparative Biochemistry and Physiology*, **40A**, 495–501.

Zeuner, F.E. (1959). *The Pleistocene Period: its climate, chronology and faunal succession*. London: Hutchinson & Co.

Zeuner, F.E. (1963). *A History of Domesticated Animals*. London: Hutchinson & Co.

Zhigunov, P.S. (ed.) (1969). *Reindeer Husbandry*. Jerusalem: Israel Program for Scientific Translations.

Zuckerman, S. (1953). The breeding seasons of mammals in captivity. *Proceedings of Zoological Society of London*, **122**, 827–950.

Østbye, E. (ed.) (1975). Hardangervida, Norway. In *Structure and Function of Tundra Ecosystems*, ed. T. Rosswall & O.W. Heal, pp. 225–264. Stockholm: Ecological Bulletin No. 20.

Index